WITHDRAWN FROM
TSC LIBRARY

George Gauld: Surveyor and Cartographer of the Gulf Coast

George Gauld
Surveyor and Cartographer of the Gulf Coast

by
John D. Ware

revised and completed by
Robert R. Rea

A University of Florida/University of South Florida Book

University Presses of Florida
Gainesville/Tampa

Copyright © 1982 by the Board of Regents of the State of Florida

Printed in the United States on acid-free paper

Library of Congress Cataloging in Publication Data

Ware, John D., 1913-1973.
　George Gauld, surveyor and cartographer of the Gulf Coast

　"A cosponsored publication of the University of Florida and the University of South Florida."
　Bibliography: p.
　Includes index.
　1. Gauld, George. 2. Surveyors — Scotland — Biography. I. Rea, Robert Right, 1922- .
II. Title.
TA533.G38W37　　526.9'9'0924　[B]　81-16341
ISBN 0-8130-0708-9　　　　　　　AACR2

　　University Presses of Florida is the central agency for scholarly publishing of the State of Florida's university system. Its offices are located at 15 Northwest 15th Street, Gainesville, FL 32603. Works published by University Presses of Florida are evaluated and selected by the faculty editorial committees of Florida's nine public universities: Florida A&M University (Tallahassee), Florida Atlantic University (Boca Raton), Florida International University (Miami), Florida State University (Tallahassee), University of Central Florida (Orlando), University of Florida (Gainesville), University of North Florida (Jacksonville), University of South Florida (Tampa), University of West Florida (Pensacola).

Contents

	PAGE
LIST OF ILLUSTRATIONS	vii
JOHN D. WARE: AN APPRECIATION, by Samuel Proctor	ix
PREFACE	xi
INTRODUCTION	xv

CHAPTER

I.	The Early Years	1
II.	Voyager to West Florida, 1764	13
III.	George Gauld's Pensacola	28
IV.	Espiritu Santo in East Florida, 1765	43
V.	Coastal Surveying, 1765–1766	58
VI.	Coastal Surveying, 1766–1767	80
VII.	Surveys to the West, 1768	95
VIII.	Service Ashore and at Sea, 1769–1771	115
IX.	Jamaica to Key West, 1772–1773	140
X.	Exploring the Limits and Beyond, 1774–1778	159
XI.	The Surveyor at Home	186
XII.	The Impact of the American Revolution	202
XIII.	Corpus Cartarum Vivarum	221

APPENDIX

A Note on the Sources and a Checklist of the Works of George Gauld	235
INDEX	243

List of Illustrations

	PAGE
I. "A View of Pensacola in West Florida" ... FOLLOWING	60
II & III. Key West and the Dry Tortugas	"
IV. "A Survey of the Bay of Pensacola 1766"	"
V. "A Survey of the Bay of Espiritu Santo in East Florida" ...... FOLLOWING	92
VI, VII & VIII. The Coast of the Gulf of Mexico	"
IX. "A General Description ... of West Florida"	131
X. "A Plan of the Mouths of the Mississippi" FOLLOWING	156
XI. The Caribbean and Gulf of Mexico	"
XII. *A Chart of the Bay and Harbour of Pensacola* ...... FOLLOWING	188
XIII. "A General Plan of the Harbours of Port Royal and Kingston Jamaica"	"

John D. Ware: An Appreciation

It is said that some men are born with salt in their blood. It must have been so for John Ware, whose family had been going to sea for four generations before him. His great-grandfather Charles operated a schooner on Chesapeake Bay, and his grandfather, Lambert Milbank Ware, sailed into Saint Andrew Bay, Florida, about 1879 aboard the *Gedney,* a U.S. Coast and Geodetic Survey vessel. He returned to the bay the next year sailing his own sloop, the *Hope.* Later Lambert Ware's schooners carried passengers and hauled freight — salt, fish, lumber, and live gophers (land turtles) — from Apalachicola to Pensacola and New Orleans. John Ware's father, Otway, was a bar pilot on Saint Andrew Bay for many years.

John Ware joined the merchant marine after he was graduated from high shool in 1932 and served his apprenticeship on Saint Andrew Bay. During World War II he became a deck officer in the U.S. merchant marine. He was a seafarer and shipmaster with Waterman Steamship Corporation at the time he was appointed a Tampa Bay pilot in 1952. He was always recognized as one of the best pilots in the business. He held the United States Coast Guard license as master of steam and motor vessels, and he piloted ships along the Florida and Gulf coasts, the Mississippi River, and several bays and harbors on the east coast of North America. Ware was secretary of the Florida State Pilots Association and a member of the Tampa Bay Pilots Association.

Captain Ware also was held in high repute as a historian of eighteenth-century Florida. His preferred research areas, the first and second Spanish periods of Florida's history, demand a special paleographic skill. To use the wealth of data contained in contemporaneous Spanish documents, one must be able to read a complex and difficult script. Though never formally trained as a paleographer, Captain Ware, by persistent effort, became a skilled researcher. No detail was ever too small to escape his attention as he examined primary source records in the archives and libraries of Spain, Portugal, and the United States. His articles were published in the *Florida Historical Quarterly,* in the St. Augustine Historical Society's journal, *El Escribano,* and in *Tequesta,* published by the Historical Association of Southern Florida.

John D. Ware: An Appreciation

He wrote the introduction and compiled the index to P. Lee Phillips' volume *Notes on the Life and Works of Bernard Romans* in the Bicentennial Floridiana Facsimile Series, published by University Presses of Florida for the American Revolution Bicentennial Commission of Florida.

For many years Captain Ware was interested in the life and career of George Gauld, the eighteenth-century British surveyor and cartographer who worked along the same area of the Gulf Coast where John Ware was born and grew up. Ware was in the midst of researching and writing a book on Gauld when, on the eve of his departure for a research trip to work in the libraries of England and Scotland, a fatal cancer was detected. He continued to work on the book to the very end of his life in January 1974, but the book was left incomplete.

So fruitful was the work up to this point, however, that George Gauld could be seen as a major figure in the history of the charting of North American waters, and the manuscript, which Captain Ware had brought to a late stage of development, obviously merited publication. It was completed after additional research in British collections and seen through the press by Robert R. Rea.

Samuel Proctor

Preface

Historical research and writing may be compared to the quest for the bits and pieces of an ancient mosaic and its final restoration. One begins with the first available fragments and hopes they are significant to the whole. The more one searches and assembles, the clearer the image becomes, each segment contributing to the other, until at last it is finished or, all known sources exhausted, it defies completion.

So it has been in researching and writing about the life and work of George Gauld, a Scot employed as a coastal surveyor by the British Admiralty during the seventh and eighth decades of the eighteenth century. A relatively obscure figure, Gauld left a little-explored though well-marked trail of professional accomplishments in the New World. There remain, however, only fragmentary accounts of his activities, and apart from parish records attesting to his birth and baptism there appears to be no information available relating to his youth. His college records are somewhat more revealing but far from complete. Had it not been for the interest and evident admiration of the "Royal Geographer" William Faden, whose posthumous publication of part of Gauld's surveys includes certain biographical information, virtually nothing would have been known of George Gauld the man. His numerous official communications, most of which were letters of transmittal for his cartographic work, contain little of a personal nature. His literary efforts, both published and in manuscript, consist of geographic descriptions and sailing directions. Together with his charts, maps, and manuscript drawings, preserved in various archives and repositories but long since outmoded by their modern counterparts, they constitute a valuable source of historical geography.

This study could not have been undertaken without the assistance of many individuals and the cooperation of numerous institutions. The writer is deeply indebted to Jack D. L. Holmes, formerly professor of history at the University of Alabama in Birmingham, who not only relinquished his claim to Gauld after years of interest and much research but unselfishly donated his substantial collection of

Preface

notes as well. Samuel Proctor, professor of history at the University of Florida and editor of the *Florida Historical Quarterly*, offered much-needed initial encouragement; both he and William M. Goza, past president of the Florida Historical Society, contributed materially by their critical reading of the manuscript. Research was made immeasurably easier by the gracious and knowledgeable assistance of Miss Elizabeth Alexander and her staff at the P. K. Yonge Library of Florida History at Gainesville and the kind offices of James A. Servies, director of libraries of the University of West Florida at Pensacola. Earle Newton, executive director of the Historic Pensacola Preservation Board, and Mrs. Linda Ellsworth, assistant historian, graciously contributed valuable material preserved in their files.

Walter Ristow, chief, Geography and Map Division of the Library of Congress; Whitfield J. Bell, librarian of the American Philosophical Society, Philadelphia; and Douglas W. Marshall, head, Map and Newspaper Division, William L. Clements Library, University of Michigan, were most helpful and generous in supplying needed information and copies of manuscripts in their respective institutions.

From abroad came valuable and essential research material and information. Miss Jacqueline Welds of the West India Reference Library, Kingston, and Clinton V. Black of the Jamaica Archives in nearby Spanish Town graciously cooperated on behalf of their respective institutions. Michael Godfrey, lieutenant commander, R.N. (Ret.), retired from a research post in the Public Record Office in London; Ronald Hope, director, Seafarers' Education Service and College of the Sea, London; Colin McLaren, archivist and keeper of manuscripts of King's College, Aberdeen, Scotland; and Mrs. H. Bain of the General Register Office for Scotland, Edinburgh—all were most helpful in providing material. To these must be added the names of Miss M. J. Perry, who graciously cooperated on behalf of the Hydrographic Department of the Ministry of Defence, Taunton, and J. C. Croomer, lieutenant commander, R.N. (Ret.), Bath. To the foregoing individuals and institutions this writer acknowledges a debt of gratitude which can only be partially repaid.

And finally, the writer reserves his special thanks for his beloved wife, Marge, without whose infinite patience and understanding this effort would never have been undertaken, much less completed.

John D. Ware

Preface

* * *

This is John Ware's book. It is not exactly as he left it, but I hope it is a book he would have enjoyed reading, just as he enjoyed poring over the maps and charts upon which it is based. The research is largely his. His correspondence shows the attention with which he pursued each lead, the gentle optimism with which he urged his British researchers to look again for a missing document, the patience with which he awaited the occasionally mislaid chart; his notes display the thoroughness with which he read and studied every piece of evidence. There was little that he missed. My own research, which has chanced to touch the same men and ships and beaches, has simply supplemented his and fleshed out the narrative. At the time of his death, Captain Ware had carried his account of the surveys to 1770, roughly two-thirds of the present volume, but in one way or another he traced the whole of Gauld's career. As I have reordered and rewritten much of the material incorporated in his original manuscript, it would be difficult to separate our work.

My personal acquaintance with John Ware was slight, but poring through another man's notes and sources, reviewing and revising his conclusions upon occasion, and becoming involved with his subject almost as deeply as he was creates a strangely wonderful sense of intimacy. As I studied the materials he had gathered and organized, the logs and journals and massive charts, I felt like one of Schoolmaster Gauld's midshipmen just learning the art of navigation. By the time we reached the coast of Florida, I understood the young lieutenant's gratitude for the watchful eye of the sailing-master, a role John Ware fulfilled in every way. At last I came to command the ship and must take responsibility for where she sails. The seas were clearly charted, but the course is mine.

In addition to sharing many of the scholarly debts acknowledged by John Ware, I am happy to recognize my particular obligations to Dr. Samuel Proctor, whose friendship brought us all together, and to Mrs. John D. Ware, whose warm hospitality and generosity made my task not only possible but most pleasurable. I am particularly appreciative of the cooperation of Christopher Terrell, Department of Navigation, the National Maritime Museum, Greenwich, and David Langmead, Hydrographic Department, Ministry of Defence, Taun-

Preface

ton, who made Gauld's printed and manuscript charts available to me.

Quotations from Crown-copyright material in the Public Record Office appear by permission of the Controller of Her Majesty's Stationery Office.

Robert R. Rea

Introduction

Few if any of the early coastal surveyors of the Floridas and the Gulf of Mexico left a more impressive and accurate array of charts, maps, and plans than did George Gauld, and none played so large a part in the life of Britain's southernmost American colonies. Yet his fame suffered from his modesty; his relatively early death left most of his work to be published posthumously, and his achievements redounded to the profit of other men.

Born and educated in Scotland, George Gauld was first employed by the Admiralty as a schoolmaster and only later as a coastal surveyor, an appointment extending from 1764 until 1781. Except for an eighteen-month period spent surveying and charting Port Royal and Kingston harbors in Jamaica, Grand Cayman Island, and the west end of Cuba, he devoted these years to the coasts and waterways of British East and West Florida and—somewhat surreptitiously—Spanish Louisiana. Gauld's work in the Gulf of Mexico and the Straits of Florida embraced the coastline of present-day Florida from tiny Rodriguez Key, lying near the southern end of Key Largo, along the coasts of Alabama, Mississippi, Louisiana, and Texas to as far west as Galveston Bay. His surveys were not limited to the outer coasts; he also explored their bays, sounds, and rivers. Gauld had originally intended to survey and chart the eastern seacoast and the shoals of the Florida peninsula as far north as Cape Canaveral, then to pass over to the Bahamas for a similar examination of the western side of those islands; but because of the increased activity of American privateers in that area in 1776, he considered it prudent to break off his work in the Florida Keys and proceed to the safer waters of the western Gulf of Mexico. Thus, through no fault of his own, he fell short of his goal in East Florida, but he exceeded it by some hundred leagues westward of the boundary of British West Florida.

George Gauld's work was in direct response to the demand for reliable and up-to-date charts created by Britain's victories in the Seven Years War and her extensive acquisition of American territory by the Peace of Paris in 1763. The great war for empire, familiar to Americans as the French and Indian War (1756–1763), pitted Great

Introduction

Britain and her continental ally Prussia against the power of France, Austria, and Russia. After a series of initial defeats, Britain rallied her strength under the leadership of the great war minister William Pitt, and in 1759 the Royal Navy swept the French battle fleets from the seas; Quebec, the citadel of New France, fell to British arms, and the French overseas empire lay helpless, waiting to be plucked. Spain was lured into the war in 1762 by desperate French diplomacy, only to see her prized possessions, Havana and Manila, fall to the conquering British. In an effort to strengthen Bourbon family ties, France gave to Spain all of Louisiana west of the Mississippi River, and in the peace negotiations she saved for Spain the city of New Orleans and that adjacent territory demarked by lakes and rivers and known as the Isle of Orleans. But France was forced to withdraw from the American continent, and, in order to recover Havana, Spain was forced to cede to Britain her colony of Florida. Britain thereby held, in 1763, all of North America east of the Mississippi, save the Isle of Orleans, and across that mighty river stood only Spain — or would, when she took possession of her new colony some years later.[1]

By royal proclamation, on October 7, 1763, the British government divided the newly acquired southern territory into two provinces: East and West Florida. East Florida consisted of the peninsula and the Gulf Coast lands west to the Apalachicola River. West Florida was bounded on the east by that river, on the south by Lakes Pontchartrain and Maurepas and the Iberville River, on the west by the Mississippi, and, in due course, by a northern parallel drawn from the confluence of the Mississippi and Yazoo rivers. St. Augustine, founded almost two centuries earlier by Pedro Menéndez de Avilés, was designated the capital of East Florida; Spanish Pensacola was preferred over French Mobile as the seat of government for West Florida.[2]

In the process of delineating the new colonies, establishing their institutions, and selecting their royal governors — James Grant for East Florida and George Johnstone for West Florida — the govern-

---

1. Lawrence Henry Gipson, *The British Empire Before the American Revolution* (New York, v.d.), vols. 6–8; J. Leitch Wright, *Anglo-Spanish Rivalry in North America* (Athens, Ga., 1971); Robert R. Rea, "British West Florida: Stepchild of Diplomacy," in *Eighteenth-Century Florida and Its Borderlands*, ed. Samuel Proctor (Gainesville, Fla., 1975), pp. 62–68.

2. Charles L. Mowat, *East Florida as a British Province 1763–1784* (reprint ed., Gainesville, Fla., 1964); Cecil Johnson, *British West Florida 1763–1783* (New Haven, 1943).

Introduction

ment of George III became keenly aware of the paucity of information regarding the Floridas. No adequate maps or accurate charts of the Florida coasts were to be found. Complete and thorough surveys of the new provinces were required, and toward that end the best efforts of the civil, military, and naval branches would be required. As the sea provided most ready access, it was clear that accurate naval charts of the Floridas' waters must be given a high priority.[3]

Most of the existing navigational instruments necessary for that purpose were far from satisfactory, for certain technical problems had yet to be solved before marine surveying could provide accurate and reliable information for proper charts. Imaginative artistry aside, all available charts were marred by serious shortcomings. Insofar as their flaws involved longitude they were relatively unimportant before the late eighteenth century, for seafarers of an earlier era were simply unable to determine that coordinate on the earth's surface. Navigation and position finding (or more properly position keeping) consisted chiefly of dead reckoning and parallel sailing, but for centuries navigators had been able to determine latitude with a fair degree of accuracy by observing the meridian altitude of the sun or by observation of Polaris with the cross-staff or improved backstaff.[4]

France took the lead in developing the scientific knowledge and instruments necessary for the accurate surveying from which more reliable maps and navigation charts eventually resulted. The first step in this direction came in 1666 with the founding of the Académie Royale des Sciences. The Académie had as an avowed purpose the correction and improvement of maps and sailing charts. It recognized that the solution to certain problems lay in the further study and application of astronomy, and to this end astronomical observations and conferences were begun by a select group of Académiciens. Among other things, these learned gentlemen set out to develop more accurate instruments, to improve the pendulum clock, and to establish the linear value of a degree of arc. Consistent with their objective, the first priority of the Académie was an accurate method of de-

---

3. On the state of British knowledge of the Floridas see Louis De Vorsey, Jr., "De Brahm's East Florida on the Eve of Revolution," in *Eighteenth-Century Florida and Its Borderlands*, pp. 78–82.

4. On navigational developments see Lloyd A. Brown, *The Story of Maps* (New York, 1949), pp. 212–40; Nathaniel Bowditch, *American Practical Navigator* (Washington, D.C., 1962), pp. 41–47, 53–57.

Introduction

termining longitude. The most promising technique, and that used for years by astronomers to establish the longitude of an ever increasing number of positions ashore, involved observations of the eclipses of Jupiter's Galilean satellites. The data for this method were developed and tabulated after years of observation and calculation by Giovanni Domenico Cassini, an Italian scientist working in Paris. This means of determining longitude, though reasonably accurate, required the use of a powerful telescope and a reliable pendulum clock set to the reference of some prime meridian, instruments which effectively ruled it out for navigational purposes.

The British government also realized the importance of accurately determining longitude at sea and, spurred by a series of catastrophic losses of ships and men, established the Board of Longitude in 1714, empowering it to offer a handsome reward to the person who could solve this problem. The prize was twenty thousand pounds — a fortune in those days — to whoever could determine longitude within thirty minutes of arc (equivalent to two minutes of time) on a voyage to the West Indies. After many and varied proposals had been submitted, none proving satisfactory, the board recognized that the solution lay in finding an accurate and reliable timepiece for shipboard use.

John Harrison, a Yorkshire carpenter, after years of effort and bitter disappointment, finally perfected the requisite timepiece. His Chronometer Number Four passed its final test in 1764 on the voyage of H. M. S. *Tartar* from Portsmouth to Barbados. Two years later a French clockmaker, Pierre LeRoy, constructed a chronometer which was a masterpiece of simplicity combined with efficiency, and it detracts in no way from Harrison's accomplishment that LeRoy's timepiece has since been used as the basis for all such instruments. With the invention of the reflecting octant by John Hadley earlier in the century and the development of a reliable chronometer, the tools were finally at hand for determining longitude at sea. Marine surveying and cartography would take on a new meaning and be transformed from an art to a science. The new charts, including those of George Gauld, would reflect this progress.

No one questioned the fact that Britannia ruled the waves in 1763. Her naval and merchant vessels had regularly plied the Atlantic sealanes for more than a century and a half; yet there was still much to be learned of American coasts and harbors from Maine to Florida, for despite her naval preeminence, British marine surveying and

Introduction

charting developed slowly. Early published charts, based largely on unscientific examination and therefore often erroneous and misleading, were developed from the work of Dutch cartographers, and it was not until the later seventeenth century that official charts beyond the coasts of Europe became available to merchant vessels. In 1670 John Seller published his *Atlas Maritimus*, followed shortly thereafter by the *English Pilot*. Between 1743 and 1761 more extensive editions of the latter were published. The first charts based on English surveys appeared in 1693 when Grenville Collins, a Royal Navy captain, produced his *Coasting Pilot*.[5]

These promising developments notwithstanding, British naval and merchant navigators shared a common problem: charts were distributed through private publishing firms, and even though the Admiralty commissioned many productive surveys, the availability of the resulting charts was often determined by the publishers on the basis of their anticipated sales. This situation, and the seafaring community's dependence upon the works of Seller and Collins in spite of their technical shortcomings, unquestionably provided the incentive for the publication of such later works as De Brahm's *Atlantic Pilot*, Faden's *North American Atlas*, Des Barres's *Atlantic Neptune*, and the charts, atlases, and sailing directions of less notable figures.[6]

The acquisition of Florida in 1763 brought the realization that accurate surveys and charts were completely lacking; colonial expansion and ever increasing trade gave impetus to the now possible project of accurately surveying and charting the coasts, rivers, and harbors of all the American colonies, including the Floridas. Accordingly, the Lords Commissioners for Trade and Plantations issued instructions to the colonial governors "to procure accurate Charts & Surveys of the Coasts of their Provinces, and of the principal Rivers and Harbours fit for Navigation." Assistance in the implementation of these instructions was assured by the Admiralty through Rear Admiral Colville at Halifax, who ordered all captains on the American station to cooperate in every way possible. The particular urgency of such work in the newly acquired territories was reflected in the

5. William P. Cumming, *The Southeast in Early Maps* (Chapel Hill, N.C., 1958), pp. 38–39.
6. A. H. W. Robinson, *Marine Cartography in Britain* (Leicester, 1962), chaps. 3–4; Woodbury Lowery, *Descriptive List of Maps of Spanish Possessions in the United States, 1502–1820*, ed. P. Lee Phillips (Washington, D.C., 1912), pp. 359, 366, 377, 398; P. Lee Phillips, *List of Geographic Atlases in the Library of Congress* (Washington, D.C., 1909), vol. 3; William G. De Brahm, *The Atlantic Pilot*, introduction by Louis De Vorsey, Jr. (Gainesville, Fla., 1974).

Introduction

Board of Trade's special orders, transmitted by Colville to the captains stationed on the coasts of the Floridas:

the Lords Commissioners of Trade and Plantations having instructed the Governours of both Provinces of Florida to explore and investigate, with great Care and attention, the hitherto unknown Coast of that extensive Country, and to procure accurate Charts and Surveys of the said Coast, and of the principal Rivers and Harbours fit for Navigation, you are to cooperate with and assist the Persons employed by the said Governours therein, as far as may be consistent with the other Services you are employed on.[7]

The accumulation of geographic and hydrographic knowledge of the older colonies, much of it of a general and unreliable nature, would be increased both in quantity and quality by these orders. Apart from a few well-known harbors and rivers, knowledge of the vast territory of the two Floridas was far from reliable. Even contemporary Spanish maps, some drawn as late as 1768, showed the southern part of the Florida peninsula as an archipelago.[8] English cartographers inevitably tended to perpetuate such errors until George Gauld and his fellow surveyors began their work.

    7. Colville to captains under his command, October 15, 1763: ADM 1/482.
    8. An example of the deterioration of Spanish geographic knowledge may be seen in Juan Josef de la Puente, *Descripción Geográfica de la parte que los Españoles poseen actualmente en el continente de la Florida* (c. 1765), Museo Naval, Madrid.

CHAPTER I

The Early Years

A plain Portland stone in the burying ground of the chapel in Tottenham Court Road, London, marked the end of the saga. The inscription read: "GEORGE GAULD, A.M. Surveyor of the Coasts of Florida, &c. . . . born at Ardbrack in the parish of Botriphny, Bamffshire, North Britain; and educated at King's College, Aberdeen . . . He died at London the 8th June 1782, Aged 50 years."[1] The beginning lies in parish records which disclose that on October 26, 1732, "William Gauld in Eastertown had a Child by his Spouse Janet Moir baptized this day called George. The Revd. Mr. George Campbel, Minr. George Cobban, Janet Gauld & Elspet Mitchel witnesses."[2]

Ardbrack is no longer to be found in the list of Scottish place-names, but the little parish church of Botriphnie still nestles in the wooded valley of the River Isla, midway between Dufftown and Keith, its yard now quiet beneath the sheltering hills. Far up the slope one gray granite house marks the site of Eastertown. Across the glen, green hills meet a blue sky where dancing clouds mock the grazing sheep in nearby fields. Central Banffshire is a proud land giving rise to sparkling streams that lend their names to noble whiskies. Its beauty enchants the eye and cloaks the sudden onslaught of freezing rain and sleet, even on a late summer day. It is a country that has bred strong men, enduring men who traveled far from their hillside homes and often by sea, for the rocky shores of Spey lie but a long day's walk to the north. Such a man was George Gauld, whose path led to the coast of Florida.

No account of young George's childhood survives, but fond memories of his native land must have inspired him in later years to give the homely names of Scottish streams — Isla, Fiddich, Dullen — to the lazy rivers of West Florida. His clear, firm signature as a member

---

1. *An Account of The Surveys of Florida* (London, 1790), p. 27. George Whitefield, the great preacher of the evangelical revival in Britain and America, opened the chapel in Tottenham Court Road in 1756. The site was devastated by bombing during World War II, and only a token marker now commemorates the original graveyard.

2. Register of Births and Baptisms, Parish of Botriphnie, County of Banff, I, 17, 3d entry: Gauld and Moir. General Register Office, Edinburgh.

"in the first class" of his college, identical with that to be found on later documents, attests to the quality of schooling the lad enjoyed. At the age of eighteen, Gauld entered King's College in old Aberdeen. He was the beneficiary of the bursary or scholarship founded by the Rev. William Watson in 1699, which provided him the modest sum of twenty pounds Scots (£1.8.0 Sterling) from one Michaelmas to the next through the academic years 1750–51, 1751–52, 1752–53.

Founded in 1494, King's College was preeminently the Highlander's university. A small, impoverished institution in the mid-eighteenth century, it was only belatedly affected by those developments that have been called the Scottish Enlightenment. In 1753 King's College adopted new regulations and a new course of study, and Gauld must have benefited from the innovative spirit. Although conservative enough to require students to reside at the college, the revised curriculum included Greek, natural history, natural and experimental philosophy, and particularly emphasized mathematics; Gauld's enrollment in the Greek class of 1751, a taste for natural history (evidenced by his membership in the American Philosophical Society in later years), and his knowledge of French and Spanish all mark him as a true scholar with a wide range of intellectual interests.[3]

He did not take his degree at the end of his years as an undergraduate at King's, but this was not an uncommon practice at the time. The actual bestowal of a degree was costly and the student body uniformly poor. Having secured a good education, George Gauld may well have had to forego—for a period—the academic trimmings. In fact, Gauld disappears from sight between 1754 and 1757, when he accepted appointment as a schoolmaster in the Royal Navy. His experience as schoolmaster may have convinced him of the desirability of a degree and probably provided him with the requisite funds to secure it. He wasted no time, once he was free of his first assignment aboard ship, in returning to Aberdeen for that purpose. The minutes of King's College and University for August 25, 1759, record that "The said day the University conferred the degree of Mr. [Master] of Arts upon George Gauld teacher of Mathematics aboard the Preston Man of War & signed a Diploma for that Effect."[4]

---

3. The King's College archives preserve in MS K. 27 receipts signed by George Gauld for the academic years 1750–53; MS K. 30 refers to Gauld and the Watson bursary; MS K. 15, p. 187, attests to his enrollment in the Greek class of 1751. See also P. J. Anderson, ed., *Roll of alumni in Arts of University and King's College of Aberdeen 1596–1860* (Aberdeen, 1900), p. 194.

4. Minutes of King's College and University, MS K. 44, p. 54; P. J. Anderson, ed., *Officers*

G^{eo.} Gauld — Cartographer

The new laureate had already put his knowledge to work in the service with which he would be associated for the rest of his life.

The position of schoolmaster aboard ships of the Royal Navy had been created by an Order in Council of 1702. It was a civilian appointment, not a naval rank or rating, and it carried the lowly stipend of twenty-four shillings per month, the bottom of the pay scale in the civil branch. A young college graduate was offered, as financial inducement to enter naval employment, the same salary as that paid to a fourteen-year-old midshipman ordinary. This undoubtedly accounted for the program's singular lack of success in its early stages. The devoted or desperate scholar who accepted appointment as schoolmaster was charged with the general educational oversight of a dozen or so teenage midshipmen aboard his vessel. He was particularly responsible for teaching them the mathematics necessary for navigation, while the captain or master supervised their practical training in the arts of seamanship.[5]

When Gauld began his career as schoolmaster, the Royal Navy was vigorously engaged in war with France and had been since 1755. The Seven Years War would produce some of Britain's greatest fleet actions and proudest victories, but the most common duty of His Majesty's ships lay in the protection of British merchant vessels and the destruction of the enemy's commerce. A fifty-gun ship such as *Preston*, on which Gauld first served, was too small to claim a place in the line of battle; convoy assignment was nearly all that she could expect in European waters.

Neither salary nor excitement seems likely to have drawn Gauld to the navy, although he may well have thought it properly patriotic to put his scholarly competence at the disposal of the senior service in time of war. Some thirty years later, Gauld's publisher suggested that another reason led the young Scot to seek a berth as schoolmaster. "Mr. Gauld embraced that," wrote William Faden, "as the only opportunity he ever had of treading on classic ground."[6] For George

---

and Graduates of University and King's College Aberdeen 1495–1860 (Aberdeen, 1893), p. 241; Anderson, *Roll of alumni in Arts*, p. 194, Appendix A, "M.A.s not alumni."

5. Michael Lewis, *The Navy of Britain* (London, 1948), pp. 192–95; Lewis, *A Social History of the Navy, 1793–1815* (London, 1960), pp. 257–61; F. B. Sullivan, "The Naval Schoolmaster during the eighteenth century and the early nineteenth century," *Mariner's Mirror* 62 (1976): 311–26. A bonus of £20 was subsequently added to the schoolmaster's annual income, and allowance was made for the size of his ship and the number of his students. Gauld probably received about £50 a year in all.

6. *Account of The Surveys*, p. 5.

The Early Years

Gauld, the navy was a substitute for the Grand Tour. If wanderlust and curiosity concerning distant and ancient lands explain Gauld's decision, both would be amply satisfied by the navy in the years to come.

George Gauld's connection with the navy began officially on July 6, 1757, when he was certified as schoolmaster; nine days later he was warranted to H. M. S. *Preston*, Captain John Evans. His ship had been anchored for two weeks in the Downs, hard by the town of Deal on the Straits of Dover, and was taking on stores in preparation for her first cruise under Evans's command. Schoolmaster Gauld began to instruct his young charges on July 15 when *Preston* set sail for the Flemish port of Ostend.[7]

The short passage out of the Downs, south of the treacherous Goodwin Sands, and across to the Flemish coast was made in a matter of hours.[8] *Preston* found anchorage just three miles from Ostend, but a terse notation in her logbook for Monday, July 18, 1757, reflects the futility of her mission: "Ye cutter returned with the account that ye French were Entering ye Town of Ostend to take possession." Reacting to this news, Captain Evans ordered the decks cleared, and exercises with the great guns and small arms were held during the next three days; but Ostend was lost, and its usefulness to the British was at an end. On July 23, orders were issued to heave anchor and return to the Downs.

Preston's next voyage took her on patrol along the east coast of England as far as Great Yarmouth, where she anchored for eight days. On August 11 Evans renewed his patrol. Fresh northerly gales and hard squalls in the North Sea compelled the captain and his sailing master[9] to lower the topgallant yards and set up the topmast shrouds, but a heavy squall soon split and blew away the foretopsail. During succeeding days several strange vessels were sighted; decks were cleared for action and battle stations manned, but all proved to be

---

7. Certificate as schoolmaster, July 6, 1757; warrant to *Preston*, July 15, 1757: ADM 6/18. Also see List of chaplains, masters at arms, and schoolmasters: ADM 6/185.

8. All subsequent information regarding *Preston*'s voyages is taken from the master's logbooks (ADM 52/986) and captain's log (ADM 51/746). Captain Ware plotted these voyages on U.S. Defense Mapping Agency Hydrographic Center Chart N.O. 37010 (formerly H.O. 4841), *The North Sea, Southern Part*, 13th ed., 1947; Chart N.O. 35000, *British Isles*; Chart N.O. 310 (formerly H.O. 4300), *Mediterranean Sea*, 17th ed., 1942.

9. The distinction between captain and master is discussed in Lewis, *Navy of Britain*, pp. 162–75. Briefly stated, the captain was the fighting man in command; the master was the seaman in all but complete control of sailing and navigating a ship.

neutral Dutchmen. In due course *Preston* crossed the North Sea, and on August 20 she stood off the island of Heligoland, where Evans deemed it prudent to anchor to avoid being blown on a lee shore by the strong westerly gales sweeping across the North Sea. On September 2, the foul weather moderated; *Preston* came to sail and proceeded to the mouth of the Elbe River, anchoring just a mile from the town of Cuxhaven. For a month *Preston* lay there, along with many other British ships, awaiting the outcome of events on shore.

Beyond the range of the navy's guns, the fortunes of war favored France. Anglo-Hanoverian troops under the command of the Duke of Cumberland were defeated by the French, but the Convention of Klosterseven allowed Cumberland to remove his army from the Continent, and the assembled vessels in the Elbe estuary were given that task. On October 4, *Preston*'s boat came alongside and delivered the baggage of the unfortunate British commander. Two days later Cumberland himself came aboard with a royal twenty-one gun salute ringing in his ears, but he remained only overnight. The following day he transferred to another vessel, leaving his personal effects aboard *Preston*. The assembled convoy sailed for England on October 7 and anchored off Deal on the twelfth. The duke's baggage was sent ashore, fresh stores were received, and the next day *Preston* sailed for Spithead and two months of overhauling and refitting. Schoolmaster Gauld's first six months in the navy had not been uneventful; he had encountered no enemy at sea, but he had seen what his forces could accomplish on land. He had seen foul weather and a foul-tempered royal duke, and he had doubtless taken advantage of the long periods of idleness that were so much a part of naval life to advance his students' knowledge and competence.

Finally, on December 18, 1757, all repairs completed and stores aboard, *Preston* left Portsmouth for Falmouth, where she would pick up a convoy bound for the Mediterranean. At noon on January 16, 1758, *Preston* and fifteen merchant vessels stood out of Falmouth harbor for the open sea. At last George Gauld was on his way to the Mediterranean where he might truly "tread on classic ground."

The fifteen-day passage to the Straits of Gibraltar was marred by hard gales and heavy seas; by the time *Preston* reached Cape St. Vincent, the southwest tip of Portugal, the convoy had dwindled to four ships. Evans proceeded slowly southward, and off the entrance to the straits he came up with Admiral Osborne's fleet and passed under his command. A period of almost two weeks was spent patrolling the At-

5

lantic approach to the straits and waiting for *Preston*'s stragglers. On February 10, the signal was finally made to proceed through the straits, and by noon the next day *Preston* was standing in to Gibraltar Bay. The twenty-seven-day passage had cost *Preston* six crewmen and one marine, but George Gauld's first sight of Gibraltar and Mount Acho on the African coast, the Pillars of Hercules which marked the westernmost limits of navigation for ancient mariners, must have wiped from his thoughts the ever present dangers of life at sea.

The fleet remained at anchor at Gibraltar until February 13, when it began a series of patrols designed to sweep the western end of the Mediterranean of enemy shipping. One cruise took the fleet close along the coast of Spain, past Malaga and Cartagena, and as far east as Cape Palos. There, on March 1, Gauld in *Preston* observed the capture of the sixty-four-gun French ship *Orpheus* by *Revenge* and *Berwick*. Twenty French prisoners were sent aboard *Preston*, and her crew was placed on short bread rations because of the additional mouths to be fed. On March 16, Evans received orders to proceed independently to Gibraltar and anchored there on the twenty-seventh.

For the next several days *Preston* was engaged in taking on stores and water for a four-month cruise, and Captain Evans was busy instructing the masters of the nine merchantmen he was ordered to convoy to various northern Italian ports. All took reassurance from the safe arrival of a large convoy from the eastern Mediterranean, and shortly after dawn, April 9, Evans gave the signal to heave anchor and make sail.

Preston set course for the relative security of the North African shore, and the second day out of Gibraltar her convoy stood off Cape Tres Forcas. The merchantmen were of different types and rigs, but their speed had been a creditable five knots under fair winds and currents. Proceeding along the Barbary Coast, Evans deemed it prudent to take certain precautions. Although it was unlikely that pirates would molest a convoy protected by a man-of-war, he gave orders for the exercise of the guns and signaled the merchantmen to gather astern of and under the protection of his ship. Remaining close to the African shore, the convoy passed Cape Ferrat (Pointe de l'Aquille) and Cape Ivi, where the first merchantman left the group. Cape Ténès soon fell astern, and a northeasterly course was set for the seaport republic of Genoa. On the morning of April 27 Evans was within four or five leagues of Genoa, and five of the convoy bore away for the harbor entrance.

With the three merchantmen remaining, Evans now set a southeasterly course for Leghorn (Livorno), but because of light winds and calms, the better part of three days was required to make the seventy-five-mile passage. Arriving off Leghorn in a calm, the English ships launched their boats and were towed into the anchorage in midafternoon, May 1.

Here *Preston* remained for twenty-two days while the crew overhauled the rigging and sails and the ship was painted from top to bottom. Fresh water and provisions, including beef and two live bullocks, were taken aboard. The British consul visited the ship, receiving an eleven-gun salute, and finally on May 15, *Preston*'s crew was granted pratique — shore leave — and allowed to visit the city of Leghorn. It may be assumed that the schoolmaster was among those who took advantage of the opportunity.

On the evening of May 23, *Preston* weighed anchor and came to sail in company with H. M. S. *Prince Edward*. Proceeding to the south, with Corsica on the west and leaving the islands of Elba and Monte Cristo to the east, they made their way to the southern end of Corsica, where *Prince Edward* took its leave. Midday, June 2, some fifteen leagues north of the west end of Sicily, a vessel bound from Alexandria to Leghorn was halted, and Captain Evans learned from her master that a sixty-four-gun French man-of-war and a frigate were cruising in the vicinity of the island of Pantelleria. The position and course of *Preston* suggest that Evans intended to pass to the west of Sicily and to proceed through the broad Sicilian Straits to his destination in the eastern Mediterranean. This route would have taken *Preston* within the cruising area of the two enemy warships. Outnumbered and outgunned, Evans apparently reconsidered and decided to take the alternate route between Italy and Sicily: better to expose his vessel to the natural perils of the twin whirlpools of Scylla and Charybdis on either side of the treacherous Strait of Messina than risk an encounter with a superior force.

Accordingly, Evans bore away to the eastward, passing the islands of Ustica and Stromboli. Three leagues from the northern entrance to the strait, which the master noted in the log as "ye Vear of Messina," two guns were fired as a signal for a pilot. It was the following morning before a pilot came alongside, only to advise Evans to anchor and await a more favorable wind and current; but at dawn the next day he came aboard, and the anchor was weighed and sails set to make the short but perilous passage. Although the weather was squally, a brisk north wind carried *Preston* safely through the Strait of Messina. The pilot was soon discharged and course set for Cape Spartivento at the southernmost tip of Italy. With the advantage of forewarning, and not just a little daring, Captain Evans had eluded the French cruisers.

Under full sail and driven by a fair wind, *Preston* crossed the Ionian Sea in little more than two and a half days to a position some eight or nine leagues from Cape Matapan (Akra Tainaron) at the southern end of the Peloponnesos. She tarried here for two days and halted several vessels which were identified as neutrals and allowed to proceed. For the next week Evans cautiously worked his way between the Greek mainland (then part of the Ottoman Empire) and the Cyclades Islands to Andros, the northernmost of this group. From there he set course across the Aegean Sea for the sixty-mile passage to the south end of the island of Khios (Chios), hard by the Turkish coast, where he encountered such light airs that he was compelled to launch the small boats to tow *Preston*. At 11 P.M., June 20, the master noted in his log: "Felt a Violent Shock of an Earthquake," and three hours later he reported that they "felt another shock as ye former." Her sails still hanging limp for lack of wind, *Preston* was towed to a position near Khios Roads, where she anchored on June 21.

Preston remained in Khios Roads for the better part of six weeks while the crew was employed overhauling and repairing the sails and rigging. The vessel was heeled; the bottom and sides were payed or painted. Caulkers were employed on the decks. Scarcely a day went by that fresh beef or mutton was not brought aboard for the ship's complement, thereby allowing them to improve their diet and conserve their salt and dried provisions for the return passage. Ships of various countries arrived and departed, saluting or being saluted by *Preston*. On June 25, the French man-of-war *Tryton* (64) and frigate *Plaidas* (30) anchored in Khios Roads, leaving two days later. These were probably the two vessels Captain Evans had taken pains to avoid by sailing through the Straits of Messina. Needless to say, no salutes were exchanged between the British and French ships, but respect for the neutrality of their Turkish host also restrained them from opening fire on one another. Finally, on the morning of July 31, *Preston* heaved anchor and set sail for Smyrna (Izmir). A pilot had been engaged and accompanied Evans's vessel out of Khios Roads, providing navigational advice on the short passage through the narrow waterway between Khios and the mainland and into the Bay of Izmir and safe anchorage at Smyrna.

Preston's stay at Smyrna was a long one. Mooring directly off the town on August 6, 1758, it would be six months to the day before she left the Turkish port, and only routine duties occupied the ship's company. As was customary, firewood, water, and fresh beef and mutton were brought aboard. The English consul paid a visit — with the usual formalities. Shortly after her arrival, several sick crewmen were sent ashore for medical attention, followed soon by "6 marines . . . to be Centinels at ye hospital." Surveillance was necessary, as evidenced by the punishment of seaman Joseph Barret with two dozen lashes for attempting to escape from the infirmary, but most of the sailors who were sent ashore for treatment were returned fit for duty. Their services were needed, for twenty-eight men had succumbed during the seven months *Preston* was in the Mediterranean prior to her arrival at Smyrna. Among these were the chaplain, the Reverend Mr. South, Mr. Powel the purser, and seaman "John Wyatt [who] by accident fell from ye Main Topsail Yard down upon deck and was killed."

An examination of *Preston*'s stores revealed that during her long voyage the gunpowder had absorbed excessive moisture. This required that it be sent ashore, in three approximately equal lots, and spread under a canvas tent for drying and airing, but in less than a week the entire ninety-seven barrels had been reconditioned and returned to the ship's magazine without mishap. On October 23, a fifteen-gun salute was fired in commemoration of His Majesty's accession to the throne, and on November 10 a twenty-one-gun salute hailed his birthday; the dried and ventilated powder evidently passed these ceremonial tests.

Finally, on January 6, 1759, *Preston* was in readiness to sail. Last-minute stores of firewood, lamp oil, candles, fresh fruit, and wine had been received, and *Preston* left the Bay of Izmir in company with four merchantmen. The passage from Smyrna to *Preston*'s next anchorage, at Iskenderun in the Gulf of Alexandretta, was initially uneventful. Then, on January 17, Evans spoke the *Hawke,* a ship sailing under letters of marque and likewise bound for Iskenderun. Governed by the speed of the slowest vessel in his convoy, Evans and his charges soon fell behind the privateer; but four days later, just south

The Early Years

of Cyprus, two sails were sighted, and when Evans gave chase he found them to be an unidentified vessel with *Hawke* in hot pursuit. At 1:00 P.M. the stranger hoisted French colors and opened fire on *Hawke*. *Preston* joined the action, and soon the hopelessly outmatched *Magdalene* of Marseilles struck her flag and surrendered to *Preston*. A prize crew of a mate and twenty men was sent aboard; the French prisoners were transferred to the man-of-war, and by evening all were under way for Iskenderun. *Preston* anchored in "Scanderoon Bay" on February 3. On the twentieth, after the customary repairs and provisioning, *Preston* and an enlarged six-vessel convoy, which included the prize, stood out of the Gulf of Alexandretta.

Captain Evans's final stop, before the long return passage to Gibraltar, was at the ancient port of Salamis on the east coast of Cyprus. There he replaced the main topmast crosstree and brought aboard "men from ye Shore to Repair ye coppers," for leaky kettles in the galley would have made it sad business preparing food for the ship's company. Last minute supplies were brought aboard at Salamis before *Preston* and her convoy set sail, midmorning March 2, on the final leg of their voyage.

The arrival at Gibraltar of the man-of-war with her prize and merchant vessels marked the end of a two-thousand-mile cruise. On his homeward passage of forty-five days, Evans chose the most direct route, which took him within sight of the ancient stronghold of Malta and through the Straits of Sicily; *Preston* and her convoy averaged a respectable four and one-half knots between Cyprus and Gibraltar.

So ended fourteen months of patrol and escort duty in the Mediterranean. *Preston* had visited the cradle of Western civilization, and her sojourns at Leghorn, Khios, Smyrna, Iskenderun, and Salamis would have permitted George Gauld to satisfy his ambition "to tread on classic ground"; one can only surmise that he availed himself of every opportunity to go ashore, view the antiquities, and observe the people and culture of the places he visited. His service as schoolmaster on *Preston* was drawing to a close. On May 12, Captain Evans departed Gibraltar; on June 25 he anchored at Spithead, and on the first of July he sailed into Portsmouth harbor, where extensive refitting of *Preston* was begun. By July 26, 1759, the ship was again ready for sea. In the morning the "Commissioners Came on Bd. Pay'd ye Ship's Company to ye 31st. of December 1758." George Gauld received his pay along with the crew, but unlike them he was immediately discharged.

In fact, Gauld left *Preston* at his own desire. As soon as the ship had returned to Portsmouth, Gauld had asked to be discharged. Captain Evans forwarded his request to the Admiralty and was ordered, on July 2, to secure a written statement from his schoolmaster. Two days later Gauld responded, explaining:

that in the first place all the young Gentlemen, and others who, by your Desire, applied to me, are instructed in the several Branches required, and consequently my service can be no longer necessary: otherwise I should have been willing to continue, as the Ship is very agreeable to me. My next Reason is that I have some views of a more advantageous Employment. These, Sir, are the only Reasons I can give, and am hopeful their Lordships will approve of them.

John Evans promptly dispatched Gauld's letter to the Admiralty together with his own opinion that the schoolmaster was "a very diligent, well behaved, good Sort of Man & has always been very careful in instructing all the young Gentlemen belonging to the Ship." The Admiralty duly honored Gauld's request and on July 14 ordered him to be discharged. He remained with *Preston* through the current pay period and left the ship on July 26.

While aboard *Preston*, Gauld had taught seventeen midshipmen up to January 1759 and eight between January and July. Evidently his instruction extended to other members of the ship's company as well, perhaps junior officers and ratings seeking to improve their prospects of advancement. Like any teacher, Gauld enjoyed mixed success with his pupils: some — the brighter ones, it may be assumed — were, from time to time, transferred from *Preston* to serve on other ships. Some were found unsuitable for naval careers and so discharged. A few succumbed to the rigors and dangers of naval life, and their names on the muster roll are followed by the doleful notation "D. D."—discharged dead.[10]

Apparently George Gauld took advantage of his naval pay to return to Scotland, after leaving *Preston* in July 1759, and visited Aberdeen in order to arrange the formalities connected with his acquisition of the M. A. degree in August — little more than the payment of the requisite fees, in all probability. Whether by choice or lack of other employment, he returned to the navy, and on June 11, 1760, he was appointed schoolmaster for *Centaur*, a fine seventy-four-gun

10. Evans to Admiralty, July 4, 1759; Gauld to Evans, July 4, 1759: ADM 1/1760; Muster book, *Preston:* ADM 36/6367.

The Early Years

man-of-war recently captured from the French and now commanded by a rising young captain, Arthur Forrest, just returned from a highly successful tour of duty in the Caribbean.[11] Gauld, who was in London at the time, joined *Centaur* in the Nore on June 15 and was soon at sea on patrol duty in the Channel and the western approaches. The cruise was routine; no enemy vessels were encountered, and on August 21 *Centaur* anchored at Plymouth, where she underwent drydocking and refitting. In mid-November she sailed for Spithead, and there Gauld was discharged by Admiralty orders on December 12, 1760.[12] This is the last known record of his whereabouts for three years, a hiatus which was brought to an end by the renewal of his connection with the navy.

The two and a half years George Gauld spent as a naval schoolmaster cannot have added greatly to his prosperity, though perhaps they did enable him to take his degree; they undoubtedly proved his personal qualities, sharpened and developed his professional competence, and probably provided those interests and connections that led to his later career in the field of cartography. He certainly became inured to life at sea, a hard, demanding existence under the best of circumstances. He enjoyed and unquestionably seized the opportunity to apply his scholarly competence in mathematics to the practicalities of navigation, and he would have experienced the need for precise, accurate charts of distant and dangerous coasts, for the best of naval captains possessed only the most rudimentary charts. Even European and British shores were but crudely described, only the most obvious and deadly dangers precisely located.

These years also created the circumstances that provided an opportunity for Gauld's future cartographic work. The Royal Navy established its preeminence in all the world's oceans, and British victories won an empire that was but vaguely known in London when it was acquired by the Peace of Paris in 1763. Sea power and the maritime commerce that followed close on the heels of the men-of-war would require the best efforts of many men like George Gauld before the British could move with safety throughout their global empire.

    11. Warrant as schoolmaster: ADM 6/19; List of . . . schoolmasters: ADM 6/185; Muster book, *Centaur:* ADM 36/5225. The first two governors of West Florida, George Johnstone and John Eliot, both served under Forrest.
    12. Muster book, *Centaur:* ADM 36/5226.

CHAPTER II

Voyager to West Florida, 1764

In every man's life there are voids that the historian cannot fill. In the case of George Gauld, the years 1761, 1762, and 1763 are empty, as far as the record goes, yet they are obviously of decisive importance to his career. During this period Gauld moved from the modest and undemanding job of naval schoolmaster into the highly technical and responsible profession of naval cartographer—a personal achievement that cries for elucidation. Gauld's mathematical training certainly provided the strongest possible academic background for his future calling, and his shipboard experience would have been an invaluable supplement, but whence came the mechanical techniques, the perfection of the draftsman's art, the cartographic knowledge and competence that suddenly become evident in 1764?

Of no less interest is the question of Gauld's personal connections with men who were so placed as to offer him employment in his new profession, for the special naval appointment which restores him to visibility was surely not secured by the mere application of an ambitious nonentity. The normal functioning of eighteenth-century officialdom literally requires that in these years Gauld came to enjoy the patronage of someone greater than himself—but Gauld left no hint of his identity. Captain John Evans thought well of his schoolmaster, but Evans's honors and influence had yet to be won. In all probability Captain Arthur Forrest also appreciated Gauld's virtues, and Forrest enjoyed a well-earned fame, but the aggressive qualities that served him well at sea also alienated the Admiralty; although his service in the West Indies might have shown him the urgent need for naval surveying in that area, his recommendation would not have carried much weight at home. On the other hand, the new governor of West Florida, George Johnstone, was also a naval officer who had served in the Caribbean and had been first lieutenant aboard *Augusta* with Forrest, and Johnstone had specifically recommended that provision be made for surveying the coasts of his new province even before he was formally authorized to support such work.[1] The governor

1. Robin F. A. Fabel, "George Johnstone and the 'Thoughts Concerning Florida'—A Case of Lobbying?" *Alabama Review* 29 (1976): 168–73.

was also intimately involved with the expanding coterie of Scots who centered in London at the outset of George III's reign, and he pushed their advancement wherever possible. It is tempting to conjecture that these contacts contributed to George Gauld's appointment as a naval surveyor and cartographer.

What is certain is that British naval successes in the Caribbean and Gulf of Mexico, the seizure of Havana and subsequent schemes to carry the war against France and Spain directly to the Gulf Coast, made the Admiralty aware of the lack of reliable charts of these waters. The acquisition of the Floridas made possible the correction of this situation. At the urging of the Board of Trade, the Privy Council formally announced, on February 10, 1764, "We find ourselves under the greatest difficulties arising from the want of exact surveys of those Countries, many parts of which have never been surveyed at all, and others so imperfectly that the Charts and Maps thereof are not to be depended upon." The Treasury, Admiralty, and Secretary at War were consequently instructed to implement the work of surveying and to see that "no time should be lost." The Admiralty was particularly directed "that accurate Surveys should be made of His [Majesty's] Dominions upon the Continent of North America, and that Surveyors and other proper Officers should be appointed for that purpose." The governors of East and West Florida were directed to give attention to the work of charting the shores of their new colonies and to cooperate with the navy in every possible way. The Board of Trade instituted the General Survey of North America, and for the Gulf Coast the navy engaged the services of George Gauld — who seems to have been forward in expressing his readiness to participate in the great endeavor.[2]

Already, on January 15, 1764, the Admiralty Secretary had advised Gauld that as he had been represented to the Lords Commissioners of the Admiralty as a proper person qualified to survey coasts and harbors and had expressed an inclination thereto, he was to be received aboard H. M. S. *Tartar* (28) and employed by Captain Sir John Lindsay where and as he might find occasion. For his services, Gauld would be allowed ten shillings per day, to be paid every six months by the Navy Board.

2. James Munro and Sir Almeric W. Fitzroy, eds., *Acts of the Privy Council of England. Colonial Series. Vol. IV. A.D. 1745–1766* (London, 1911), pp. 619–20; Order in Council, February 10, 1764: ADM 1/5166; Louis De Vorsey, Jr., *De Brahm's Report of the General Survey in the Southern District of North America* (Columbia, S.C., 1971), pp. 1–6.

G^eo. Gauld — Cartographer

The new surveyor was thereupon ordered to report immediately to *Tartar* at Portsmouth, for Lindsay was preparing to sail for the West Indies.³

A fortnight later, Their Lordships directed the Navy Board to see that Captain Lindsay was supplied with the necessary surveying equipment. In addition to her normal complement of boats, *Tartar* was to be provided with a ten-oared pinnace and an extra four-oared boat. The surveyor would have at his disposal two theodolites, a plain table with sights, six surveying chains, a quadrant, a reflecting telescope for astronomical observations, and a case of pocket instruments. The drafting of charts would require a pair of proportional compasses, half a ream of imperial and half a ream of elephant paper, one ream of medium-sized paper, four reams of post paper, twelve skins of vellum, four dozen black lead pencils, and an equal number of hair pencils. For the purpose of producing charts on different scales, a camera obscura and a pantograph were provided. While he was attached to *Tartar*, George Gauld would be well equipped with the tools of his profession.⁴

Armed with his Admiralty warrant, Gauld boarded *Tartar* at Spithead on February 13, 1764. A minor embarrassment seems to have arisen when Gauld made his appearance. The Admiralty order authorizing Lindsay to receive him had been sent initially to the Deptford naval base, hard by Greenwich on the Thames, but *Tartar* had sailed for the Channel before the dispatch arrived. It had therefore to be returned and forwarded to Spithead. When it finally reached Lindsay it directed him to receive aboard one "John Gould." The misspelling of his last name was a burden the Scotsman would have to suffer all his life, but the further error regarding his Christian name

---

3. ADM 2/724. See also Admiralty to Navy Board, February 28, 1764: ADM A/2556 (NMM). Gauld was given the same salary as James Cook, who was employed in northern American waters. The future captain and discoverer of Australia was, at this date, merely a naval surveyor like George Gauld. Admiralty to Navy Board, January 4, 1764: ADM A/2555 (NMM). On the navy's independent surveying activity see Admiralty Minute Book, 1763–64: ADM 3/71.

4. Admiralty Minute Book, January 27, 1764: ADM 3/71; Admiralty to Navy Board, January 27, 1764: ADM A/2555 (NMM); Lindsay's receipt, [February 1764?]: ADM 1/2051. For comparison see the requirements set forth by De Brahm and Des Barres in Louis De Vorsey, Jr., "Hydrography: A Note on the Equipage of Eighteenth Century Survey Vessels," *Mariner's Mirror* 58 (1972): 175–76, and G. N. D. Evans, "Hydrography: A Note on Eighteenth Century Methods," ibid. 52 (1966): 248–49. Contemporary examples of several of these instruments are pictured in David P. Wheatland and Barbara Carson, *The Apparatus of Science at Harvard 1765–1800* (Cambridge, Mass., 1968).

obviously threatened both his status and his pay. Corrective steps were quickly taken, and on February 21 the Admiralty Secretary issued proper orders in the name of the real George Gauld.[5] Lindsay was instructed that:

Mr. George Gauld being appointed to proceed in the ship you command in order to be employed in surveying the coasts & harbours in the parts where she may be stationed, for which he will be allowed after the rate of ten shillings a day; you are hereby required & directed to receive him on board & cause him to be victualed as the ship's company and to employ him from time to time as there shall be opportunity in surveying coasts & harbours accordingly, and to cause him to make exact plans of the same, and to note carefully the soundings, settings of the tides & currents, etc., and at the end of every six months to give him a certificate of his having been so employed, in order to his obtaining his pay from the Navy Board.[6]

It had been intended that *Tartar* should carry another and more distinguished passenger — Governor George Johnstone; but Johnstone's preparations were far from complete, and on March 23, the Admiralty directed Lindsay to wait no longer but to proceed at once to the New World without His Excellency.[7] Consequently, *Tartar* sailed from Portsmouth for Barbados on March 28, 1764, a voyage of historic interest and importance to every seafarer and most certainly to George Gauld.

Captain Sir John Lindsay followed a time-honored sailing route which had its origin in the first voyage of Columbus. Departing from the English Channel, Lindsay sailed to Madeira and thence southward to the Canary Islands, but, unlike the first discoverer, he proceeded to even lower latitudes near the Cape Verde Islands before turning westward. This choice of tracks was based on two and a half centuries of practical experience in utilizing the natural elements rather than combating them. The southern route, though longer than a northerly course, was usually quicker; it took initial advantage of the favorable Portugal current and later fell in with the Canary and North Equatorial currents and the never failing northeast trade

---

5. Admiralty Minute Book, February 13, 1764: ADM 3/71; Admiralty to Lindsay, February 13, 1764: ADM 2/91; Admiralty to Lindsay, February 21, 1764: ADM 2/724; Muster book, *Tartar:* ADM 36/6849.

6. Admiralty to Lindsay, February 21, 1764: ADM 2/91; Admiralty to Navy Board, February 28, 1764: ADM A/2556 (NMM).

7. Admiralty Minute Book, February 21, 1764: ADM 3/71; Admiralty to Lindsay, March 23, 1764: ADM 2/724; Lindsay to Admiralty, March 16, 25, 1764: ADM 1/2051.

G^eo. Gauld — Cartographer

winds. A southerly course avoided the prevailing westerlies of the higher latitudes and the adversities of the North Atlantic drift current spawned by the Gulf Stream. Furthermore, the lower latitudes usually enjoyed better weather than the North Atlantic, especially during the winter months. Thus the southern route was by far the most advantageous.[8]

When *Tartar* arrived at Barbados, May 13, it ended a voyage that was a milestone in maritime history. One of her passengers was William Harrison, son of the Yorkshire carpenter-turned-clockmaker John Harrison. The father was then seventy-one years of age and had entrusted the final accuracy tests of his latest timepiece to his son and assistant. The tests were eminently successful, vindicating the elder Harrison's lifetime of patient, dedicated work and finally making it possible for the navigator to determine his longitude accurately at sea.[9] The immediate importance of this historic event was not lost on George Gauld, as the records of his subsequent surveys and cartography disclose.

Tartar's passage from the wind-whipped English Channel to the more hospitable lower latitudes was rough, though probably no worse than was usual for the season.[10] After twenty-four hours at sea *Tartar* had made only sixty miles of westing from Portsmouth in the face of strong southwesterly gales and heavy seas. It was therefore decided to anchor under the lee of the Bill of Portland until the weather moderated. The respite from the elements was used to set up (tighten) the rigging, which had been subjected to considerable stress by the violent wind and sea, and it was found that most of the nine-pound powder cartridges in the guns had been damaged by the heavy seas taken aboard.

The inclement weather continued unabated until the afternoon of

---

8. Samuel Eliot Morison, *Admiral of the Ocean Sea* (Boston, 1942), p. 157 and map facing p. 222; Morison discusses the other routes in *The European Discovery of America, The Northern Voyages, A.D. 500–1600* (New York, 1971). Master's log, *Tartar*, March 28 to May 13, 1764: ADM 52/1465, indicates the vessel took this route.

9. Admiralty Minute Book, February 4, 1764: ADM 3/71. The muster table of *Tartar* (ADM 36/6849) indicates that William Harrison and John Wilson, his assistant, joined the vessel at Longreach, February 13, and left in Barbados, May 13, 1764. Master's log, *Tartar*: ADM 52/1465; *An Account of The Surveys of Florida* (London, 1790), p. 9; Lloyd A. Brown, *Story of Maps* (New York, 1949), pp. 236–37.

10. The voyage from Portsmouth to Jamaica is traced from the master's log, *Tartar*: ADM 52/1465, and U.S. Naval Oceanographic Office Chart N.O. 126 (33d ed., 1942; rev., 1971; formerly H.O. 956), *North Atlantic Ocean-Northeastern Part*.

March 31, when the southwesterly gales shifted to the west and moderated. At six o'clock the next morning, preparations were made to proceed. At seven-thirty the anchor was aweigh, sails were unfurled, and a course was set for the island of Guernsey, almost directly south of Portland Bill, near the French coast. The day provided further interest for *Tartar*'s scientists, for an eclipse was noted in the log. After three days of weather ranging from calm to fresh gales ahead, *Tartar* had tacked some sixty-five miles across the English Channel and anchored in Guernsey Roads, where she stayed until April 6.

The much longer passage to Funchal in Portuguese Madeira involved beating to windward against the prevailing westerlies to clear the treacherous Isle of Ushant, a rocky sentinel off the French coast, which guarded both the Channel and the northern expanse of the Bay of Biscay. Once past this hazard, a navigator had more sea room but still had to make sufficient westing to clear its Spanish counterpart to the southwestward, Finisterre.

Sir John Lindsay's master, Robert Reid, accomplished the passage under adverse conditions requiring numerous tacks, shifting, shortening and reefing of sails, and striking and resetting the topgallant masts and sails as the wind commanded. There was always an element of risk for the seamen who performed these duties, especially in rough weather and at night. The log entry for 7 P.M., Tuesday, April 9, 1764, reports laconically, "In stowing the fore staysail, Thomas Brown, seaman, was washed overboard and drowned." A few hours later the violence of the storm carried away the rigging of the fore topmast staysail and split the sail, so that it had to be cut away.

Proceeding toward Cape Finisterre across the broad mouth of the Bay of Biscay, *Tartar* labored about a day and a half under weather even worse than that experienced in the English Channel. Reid, the master, realizing that the ship was being blown off course and toward the rock-bound coast of northern Spain, routed out his crew at three in the morning and gave the order to wear ship and bring the vessel about on the opposite tack. The additional sea room provided by retracing their course for more than a day, together with an improvement of weather, allowed Reid to come about and pass Cape Finisterre safely shortly after midnight, April 14.

With Finisterre over their port quarter, a southwesterly course was set for the Madeiras. Each day brought better weather and more favorable current; sails reefed during the storm were shaken out to take full advantage of the variable winds. During the three days' run to a position some eighty miles from Porto Santo, northernmost of the Madeiras, *Tartar* logged an average speed of better than seven and one-half knots. Fair weather allowed the crew to make the vessel shipshape and to perform other routine chores, which included the punishment of Francis Chambers, marine, with two dozen lashes for theft and neglect of duty.

The master's logbook for April 18, 1764, confirmed the accuracy of John Harrison's chronometer and his son's navigation in these words: "At 4 P.M. Mr. Harrison gave in writing to Sir John Lindsay the distance of westing the ship had run to Porto Santo by the timekeeper, which was 43 ms. [minutes of arc] difference of longitude west." With this information in hand, a perfect landfall was made at one o'clock the following morning. Proceeding with bare steerageway in light airs, *Tartar* cautiously approached the larger island of Madeira and at three-thirty that afternoon anchored in Funchal Roads. Two additional log entries provide conclusive if not final proof of the efficiency of Harrison's chronometer. These record the almost exact longitude of Porto Santo as 16°20' W of London and Funchal Roads as lying in latitude 32°33' N and longitude "by Mr. Harrison's watch 17°01' west of London."[11]

Tartar's stay in Funchal Roads was taken up with receiving provisions, water, and wine for the ship's company and stowing these essentials in the ship's hold. The wine was no doubt deposited in a more secure place aboard ship, to be doled out at the captain's pleasure. At eight o'clock Sunday morning, April 22, as the British consul left the vessel, he was honored by an eleven-gun salute, and the laborious preparations to sail began. The stream anchor was weighed and all but half a cable of the "best bower" anchor was heaved in. At ten o'clock brandy was served to the ship's company, and three hours later the remaining anchor was stowed and *Tartar* was under way.

By now the strong gales and heavy seas of the English Channel and Bay of Biscay had been left far behind; the winds were more favorable, though variable in force and direction. With this help and the ever increasing Canary Current, *Tartar* made her way southward at a steady if unspectacular five knots average for six days. At six o'clock the morning of April 25, the Island of Palma bore "SSE, distant 4 or 5 leagues."

11. Plots on U.S. Naval Oceanographic Office Chart H.O. 956A (4th ed., 1941), *North Atlantic Ocean*, confirm the accuracy of the coordinates recorded by Harrison.

Voyager to West Florida, 1764

April 29 saw *Tartar* in latitude 18° N, off the westernmost bulge of Africa, where she encountered the calms and light airs preceding the northeast trade winds. Here the helpful Canary Current merged into the vast westward-flowing North Equatorial Current. The order was therefore given to steer a westerly course and to trim sails to take full advantage of the favorable wind. This course took *Tartar* just sixty miles north of Santo Antão, the northernmost of the Cape Verde Islands. Having made her last tack on this passage before arriving in Barbados and with the wind over her starboard quarter, *Tartar* was running free, averaging better than seven knots. The final two weeks were marred by bad weather for less than one day when, near mid-Atlantic, Captain Lindsay was forced to reduce his light sails as some were carried away in brief gales and squalls.

William Harrison, basing his determination of longitude on navigation performed with the aid of his father's timepiece, advised Captain Lindsay of the ship's position prior to every critical landfall, not only at Porto Santo and Palma but at Barbados as well. He determined the ship's position with such accuracy, the day before arrival, that the captain and master were able to make a perfect landfall at Barbados. In view of all that preceded it, their anchorage in Carlisle Roads at ten o'clock Sunday morning, May 13, was anticlimactic. A later comparison ashore disclosed that the total error for seven weeks, six of them spent on the trans-Atlantic voyage, was only 38.4 seconds of time, or 9.6 minutes of longitude. Upon Harrison's return to Portsmouth more than five months later, an error of only 54 seconds was proven.[12]

There is no indication of Gauld's reaction to this remarkable demonstration, but it may safely be assumed that he was deeply impressed. His interest must have been more than just academic, for he would soon be called upon to determine geographic coordinates and, much more important, to record them on charts of his own making. That he was enlightened by his experience with Harrison and his timepiece on the westward voyage there can be no doubt; the written and cartographic records speak for themselves.

Five days were spent in Barbados. May 17 was the Queen's birthday, and at one o'clock in the afternoon a seventeen-gun salute was fired in her honor as preparations were made to sail. At 4:30 P.M. the anchor was aweigh and the vessel came to sail, bound for Jamaica.

---

12. Bowditch, *American Practical Navigator*, p. 47; Brown, *Story of Maps*, pp. 236–37.

G^{eo.} Gauld — Cartographer

Proceeding in fresh gales and hazy weather on a northerly course, *Tartar* passed St. Lucia, Martinique, and Dominica of the Lesser Antilles. Lindsay worked his ship through this chain of islands by way of the unnamed thirty-mile passage between Guadeloupe and Montserrat. At half past two the afternoon of May 20, *Tartar* anchored in Basseterre Road along the southwest shore of the tiny island of St. Christopher or St. Kitts.[13]

A somewhat mysterious entry appears in *Tartar*'s logbook for 8 P.M., May 18: "The Island of Deseada [La Desírade] by Mr. Harrison's timepiece is 60°46' west longitude of London." It might be inferred that William Harrison was still on board and using his father's chronometer, but Harrison had been discharged from *Tartar* in Barbados and had returned to Portsmouth, with Chronometer Number Four, on another vessel.[14] How, then, could his timepiece be in use aboard *Tartar*? The answer lies in the supposition that Harrison brought two chronometers aboard, the official test instrument, Number Four, and an unofficial timepiece which remained with the ship in the custody of Captain Lindsay. Gauld himself may have computed the longitude of the tiny island. Lending support to the two-chronometer theory is a letter dated almost five years later from Admiral William Parry to the Admiralty, in which Parry stated that Gauld wished Their Lordships to be reminded of their promise to send him such of the instruments as "were carried home by Sir John Lindsay." The admiral further related that Gauld had been "at the expense of a stop watch made by one of the best hands in London, which had all the essential properties of Mr. Harrison's timekeeper."[15]

At dawn of May 21, although the visibility was none too good, *Tartar* "weighed and came to sail . . . standing off and on under topsails." Their westward course from St. Kitts to Jamaica took them less than fifty miles south of the Virgin Islands and Puerto Rico and within sight of Points Beata and L'Abacou on the south side of Hispaniola. After standing off and on most of the previous night, await-

---

13. U.S. Naval Oceanographic Chart N.O. 410 (108th ed., 1949; formerly H.O. 1290), *Gulf of Mexico and Caribbean Sea* was used by John Ware to plot *Tartar*'s passage from Barbados to Jamaica.

14. *Tartar* muster table: ADM 36/6849; Brown, *Story of Maps*, pp. 236–37; Bowditch, *American Practical Navigator*, pp. 47, 1071. The actual longitude of La Desírade lighthouse is 61°01' W.

15. Parry to Admiralty, January 29, 1769: ADM 1/238.

Voyager to West Florida, 1764

ing daybreak, *Tartar* made sail at six o'clock, May 27, and entered Kingston harbor. Her thirteen-gun salute to Admiral Sir William Burnaby, commander in chief of the Jamaica station, was returned as she slowly worked her way past the battery to join the six vessels of the Royal Navy at anchor in the harbor.

Naval custom would have required Captain Lindsay to pay his respects to Admiral Sir William Burnaby and deliver his orders from the Admiralty. These directed Sir William to station Lindsay in the "Bay of Mexico." As "Senior Captain of his Majesty's ships stationed in the Gulf of Mexico," Sir John was authorized to fly the "broad pendant" of his rank upon arrival at Pensacola, thus becoming commodore of a squadron of two frigates and two sloops. His station comprised the Gulf of Mexico from the Mississippi River to Cape Florida, and his vessels were to be employed at his discretion "for guarding the coast . . . protecting and assisting garrisons and the trade of his Majesty's subjects." He was further instructed "to make such remarks and observations, with plans of the places, as their Lordships have directed."[16] George Gauld, too, might have paid his respects to Admiral Burnaby and acquainted him with the nature of his assignment.

The Admiralty order of March 23 explicitly directed Sir William "to recommend to the Commanders of His Majesty's Ships, the obtaining a perfect Knowledge of the Navigation on the Coast of Florida and other parts within the Limits of [his] command; and for making such Remarks & Observations of the Coasts, Roads, Winds, Currents, Situations of the British and other European Settlements &c., as are therein mentioned."

Sir William thought so well of the idea that he had already initiated a similar plan some five months before the arrival of *Tartar* and had dispatched "Captain [William Chaloner] Burnaby of the *Druid*, who is a good Draftsman, to the Bay of Mexico in order to take exact Plans of Pensacola and the Mobile."[17] Notwithstanding the admiral's endorsement and the familial relationship between the two, there is no record that Captain Burnaby's assignment produced any significant cartographic results.

His vessel being in all respects ready for sea, Sir John Lindsay set

16. Admiralty Minute Book, October 25, 1763, March 22, 23, 1764: ADM 3/71; Burnaby to Admiralty, April 8, June 4, 1764: ADM 1/238.

17. Admiralty Minute Book, March 23, 1764: ADM 3/71; Burnaby to Admiralty, February 2, 1764: ADM 1/238.

sail on August 6 on the final leg of the voyage which took him and George Gauld to a new province and new duties and responsibilities.

Tartar's voyage from England to the West Indies was of special interest because it offered George Gauld an opportunity to participate in a major scientific breakthrough. Anticipation of his first view of the coast on which he would pursue his own investigations would have heightened his enjoyment of the shorter passage from Port Royal to Pensacola. For this leg of the journey, the ship's company was joined by another peripatetic Scot, Colonel Lord Adam Gordon, who, having dutifully if briefly visited the troops of his 66th Regiment at Anotta Bay, Jamaica, embarked with Lindsay and Gauld on *Tartar* to begin an extensive tour of the British North American colonies.[18]

Although Gauld and Gordon were widely separated by social standing, they shared a common curiosity regarding the continent toward which they sailed. Gauld would spend seventeen laborious years carefully surveying a small part of its coastline; Gordon, on the other hand, would take little more than a year in a journey which led from Pensacola and Mobile to Montreal, from Charleston to Boston. Lord Adam had already begun a journal that discloses more than passing shrewdness of observation, and during the uneventful voyage north he would have discussed with Gauld the surveyor's plans and prospects. Of those days Gordon wrote:

We Set Sail early on Monday [August 6], *and after making the great Kaymana* [Grand Cayman Island] *and the Capes of Corientees and Cape Antonio, on the West End of Cuba, we kept across the Gulph of Florida falling in with the Tortuga Bank, and the Bay of Apalachy. — On this Bank we catched much good Fish of sorts not known in Europe, and some Dolphins.*[19]

The army officer's brief account records the high points of the voyage, but Gordon could not know how far from the optimum course *Tartar* had strayed. Captain Lindsay and Master Robert Reid had little idea where they were after rounding Cape San Antonio, the western tip of Cuba, and striking north. The master's log discloses that the daily difference of longitude, and the courses, distances and bearings, all obviously estimated after leaving Cape San Antonio, are at

18. See "Journal of an Officer who Traveled in America and the West Indies in 1764 and 1765," in *Travels in the American Colonies,* ed. Newton D. Mereness (New York, 1961).
19. Ibid., pp. 381, 392.

Voyager to West Florida, 1764

wide variance with *Tartar*'s actual noon positions as deduced from the usually accurate latitude and frequent soundings. When plotted on a modern chart the actual track was far to the east of that indicated by the estimated and erroneous coordinates, a fact which Lindsay and Reid did not discover until they sighted the low-lying land of Cape St. George at the western extremity of Apalachicola Bay. Even then Reid recorded nearby "Cape [San] Blass" as being "west 27 leagues." By any standard, they were lost! There was nothing to do but anchor and send a boat inshore to sound ahead and westward. They may well have been relieved of their confusion by information obtained from a sloop they halted for identification some hours later. In any event, at five o'clock the morning of August 18 they weighed anchor, came to sail, and by noon had worked their way around the outlying shoals of Cape San Blas and were headed for Pensacola.[20]

That they found themselves set far to the eastward of their presumed track is no reflection on Lindsay or Reid. Actually, they navigated cautiously in accordance with the customary practices of their day. Nor are the reasons for their miscalculations hard to find. The available charts were unreliable, if not misleading, and almost certainly would not have sufficiently indicated depths of water to provide any meaningful comparison with their frequent soundings. Furthermore, they could not have known that their vessel would be set sharply eastward by the incipient Gulf Stream entering the Straits of Florida or that the unpredictable currents of the eastern Gulf might set them still farther to the east even against the wind. Adopting a safe plan, they sounded hourly until they sighted the unfamiliar foreland of Cape St. George, where they anchored and then proceeded with deliberate caution.[21]

And so it was that some twenty-four hours later, after they had passed St. Joseph and St. Andrew bays, the west end of Santa Rosa Island bore north-northwest two miles. As the double land of the island and the main gave way to the broad expanse of Pensacola Bay, a long and tedious journey was drawing to a close for George Gauld. Lord Adam described their first sight of Gauld's new home: "An Island called Santa Rosa forms the mouth of the Harbour, and on it stands a trifling little Fort, where is commonly a Serjeants Guard with a Flag Staff, to give notice to the Fort of Pensacola, when any

20. Master's log, *Tartar:* ADM 52/1465.
21. Analysis of *Tartar*'s track drawn from *United States Coast Pilot 5* (U.S. Dept. of Commerce, 1967), pp. 51–52.

Vessel is coming in, it is customary for him to fire one Gun, and hoist his Colours." "The Harbour of Pensacola, or rather the Bay is magnificent, and might contain any Fleet, was it not for want of Water at the Bar." *Tartar* was forced to wait at anchor Sunday night, August 18, so that it was Monday afternoon before Lindsay dropped anchor about two miles off the little town of Pensacola.[22]

Their first impression of Pensacola from the deck of *Tartar* could hardly have been reassuring to the two Scots, and a tour of the village would have done little or nothing to improve the image. Colonel Gordon, ever the soldier, saw first the inadequacies of defense:

The Fort is an Oblong Square with a double Stockade and a very narrow Ditch dug in the Sand. Four Bastions are intended. — The Governour's is the only tolerable House in the place. It is covered with Shingles, and has a Balcony both ways up one pair of Stairs. —All the other Houses are on the ground, and covered with Palmeto Leaves. It is a very poor place, the Soil a deep white Sand for many Miles round.

After this disheartening description of the capital and its meager defenses Gordon turned to the more favorable elements of the area. He noted (rather mistakenly) that the garrison of Pensacola had an adequate and wholesome supply of fresh water and that the country abounded in "Pitch Pine," cedar, and "an infinity of Candleberry Myrtle." The heat he found "less intense than at Jamaica or the Leeward Islands," thanks to the sea breeze, but the garrison officers complained bitterly of the cold the previous winter and the cutting easterly winds from which they were barely protected by frame houses, "the Sides either plaster or bark of trees," roofed with palmetto leaves, "and Scarce a Chimney to be seen." Despite these conditions, Pensacola was "by all accounts a most healthy Situation," in contrast to Mobile, where "everybody is ill, several have died, and in the Regiment quarter'd there, they have but one Officer able to do Duty."

Gordon's observation that "at Pensacola they have no fresh Meat, but what is brought from the Country about Mobile and the Mississippi" was balanced by his reference to the abundance of fresh fish — "the chief sustenance of the Garrison" — the prevalence of wild

22. "Journal of an Officer," pp. 381–83. Lindsay advised the Admiralty that he arrived August 20; the discrepancy in dates simply reflects *Tartar*'s cautious progress to an anchorage near the town. Lindsay to Admiralty, October 31, 1764: ADM 1/2051; Admiralty Secretary to Lindsay, March 14, 1765: ADM 2/725.

game in the winter months, and the possession of hogs, goats, and poultry. With respect to domesticated animals, he noted that "nothing can be trusted out of the Stockade, since the Indians, who come frequently . . . make a custom of stealing everything that [they] will eat, or be usefull."[23]

During the months of September and October 1764, the newcomers to Pensacola would become familiar with their none-too-congenial surroundings and the somewhat peculiar society that gathered on this new imperial frontier. Although Pensacola had been a Spanish post since 1698 and had been established at its present site since 1757, after the first location on Santa Rosa Island had been devastated by a tropical hurricane, the Spanish occupants had departed in a body shortly after Lieutenant Colonel Augustin Prevost and the 60th Regiment had taken possession of the fort on August 6, 1763. The British then found Pensacola to be "a small village consisting of about one hundred huts surrounded by a stockade." Prevost reported that the woods were close around the village and a few paltry gardens were the only visible improvements — a condition he blamed upon "the insuperable laziness of the Spaniards."[24] The redcoats did little at first to improve the place, but Pensacola gradually attracted a trickle of Anglo-American merchants who sought, without notable success, to profit from trade with the Indians and with their neighbors at New Orleans. Military authority, its personnel shifting as troops were relieved or newly assigned to Pensacola, provided a semblance of government. The Deputy Quartermaster General for North America, Lieutenant Colonel James Robertson, paid a quick visit in order to report to headquarters the condition of the forts in West Florida. With him came Lieutenant Philip Pittman, a military engineer with whom Gauld would work at a later date.[25] Land speculators and ambitious traders wandered in, making the best they could of small opportunities and uncomfortable circumstances. As had *Tartar*, every ship brought a few more Britons to

---

23. "Journal of an Officer," pp. 382–84. Sir John Lindsay noted that there was much fish but few vegetables to be had: "I never saw so Sterile a Soil, every part . . . is entirely Sand." Lindsay to Admiralty, October 31, 1764: ADM 1/2051.

24. Prevost to Secretary at War, September 7, 1763: C.O. 5/582. Also see Robert R. Rea, "Pensacola under the British 1763–1781," in *Colonial Pensacola*, ed. James R. McGovern (Pensacola, 1974), pp. 57ff.

25. Robert R. Rea, "Lieutenant Colonel James Robertson's Mission to the Floridas, 1763," *Florida Historical Quarterly* 53 (1974): 33–48, and Philip Pittman, *The Present State of the European Settlements on the Mississippi* (Gainesville, 1973), pp. xi–xii.

G^(eo.) Gauld — Cartographer

populate the town, but Pensacola remained in great need of the leadership and order that only properly constituted authority could provide. Finally, in October 1764, while Lord Adam Gordon was off sightseeing in Mobile and George Gauld was engaged in his first cartographic duties, the long-awaited civil governor, accompanied by numerous officials, settlers, and their families, made his appearance; a new era in the life of British West Florida began.

CHAPTER III

George Gauld's Pensacola

With the arrival of Governor George Johnstone on October 21, 1764, the interim of military supervision in West Florida came to an end and the civil administration of the colony began. Johnstone, born in 1730, at Westerhall in Dumfriesshire, Scotland, was a younger son of the gentry, and it was natural for him to pursue a career in the navy. The record he compiled can only be described as erratic — distinguished on the one hand by acts of gallantry, marred by disputes and duels on the other. He was endowed with intelligence, energy, and courage combined with the less admirable traits of stubborn contentiousness and verbose self-righteousness. All of these characteristics were displayed during the slightly more than two years he served as governor in West Florida. They led him into ludicrous but fierce squabbles with the military and legal officers of the colony, but they also enabled him to meet the challenge and overcome many of the real difficulties involved in establishing constitutional government in the province.[1]

Johnstone soon identified his problems: the ruinous condition of the Spanish village of "Panzacola" with its dilapidated stockade and debilitated garrison, the threat of potentially hostile Indians, and the pressing need to develop that trade and commerce upon which the colony's future seemed to depend. After a short period during which he instituted the civil forms of government, installed his council, appointed minor officials, and familiarized himself with his new domain, the governor set to work creating a British Pensacola. At an evening session of the council on January 25, 1765, he advised:

that the next object for their Consideration was the proper place for Establishing a Town within this Harbour, that it would be necessary to take a View of the different Parts of it, that as Sir John Lindsay, Commodore of his Majesty's ships on this Coast was best Acquainted with the Harbour

---

1. Robin F. A. Fabel, *Bombast and Broadsides, The Lives of George Johnstone* (Troy, Ala., forthcoming), and "Governor George Johnstone of British West Florida," *Florida Historical Quarterly* 54 (1976): 497–511.

G^{eo.} Gauld — Cartographer

and as the convenience of shipping would be a principal Consideration, he had prepared a Letter to ask his assistance in that survey.[2]

The governor's proposal was quickly approved by the council and forwarded to the naval commandant.

Johnstone was undoubtedly aware that the information he sought had already been compiled. The two months between the arrival of Lindsay and Gauld aboard *Tartar* and that of the governor had not been wasted. For obvious reasons the senior captain at Pensacola wanted detailed information concerning this magnificent body of water, and his surveyor set about compiling it at the earliest possible moment. The ship's log contains no reference to surveying, for *Tartar* remained at anchor in the bay and was used as a base of operations. Gauld probably did not continue to live on board, but he remained on *Tartar*'s muster as a supernumerary "borne for victuals only." Virtually all the work of surveying the harbor was in or near sheltered waters, so the pinnace would have been placed at Gauld's disposal, crewmen assigned to his working party, and every encouragement given to bring his initial work to early fruition. By the end of October he had finished his first survey of American waters and compiled "A Plan of the Harbour of Pensacola in West-Florida Surveyed in the Year 1764 by George Gauld M.A. The Bar by Sir John Lindsay." The captain's personal contribution need not be doubted. Lindsay took his role in the Gulf Coast survey seriously; discovering that all extant plans of the harbor and shoals were "very erronious [*sic*]," he set his master, Robert Reid, to work, himself "assisted in the greatest part," and was able to transmit Reid's three-page "Remarks for [entering] Pensacola Harbour" to the Admiralty on October 31, together with Gauld's new chart.[3] The earliest of Gauld's productions survives in a handsome manuscript; colored and drawn on vellum, it

---

2. Council Minutes, January 25, 1765: C.O. 5/625.

3. Lindsay to Admiralty, October 31, 1764: ADM 1/2051; Admiralty to Lindsay, March 14, 1765: ADM 2/75. Gauld's chart in the Library of Congress, Washington, D.C., Geography and Map Division, Group RMR, appears to be the original; no copies have been found. Lindsay subsequently wrote, "Mr. Gauld the Surveyor put down in his Maps *by Order of Sir John Lindsay*, which I declare he did not do by my direction." Sir John thought it quite proper, however, for Gauld was attached to *Tartar*, and he cast no blame on the surveyor. In fact, the offending phrase is not found on any known Gauld chart, but the mere appearance of Lindsay's name enraged Admiral Sir William Burnaby. Lindsay to Admiralty, October 25, 1765: ADM 1/2051.

reflects his mastery of his craft. Oriented north and south, it lacks coordinates of latitude or longitude, but under "Remarks" Gauld noted that Pensacola lay in 30°30' N latitude. The chart's most prominent feature is a handsome compass rose, centrally located, with its thirty-two points emanating outward to the margins on the south and east sides. Shown on this rose is a magnetic variation of 4°30' E. Along the shoreline of the bay and to a lesser extent on the land areas bordering the Gulf of Mexico are depicted trees as they would be viewed from a southerly position. The resulting perspective gives the drawing a curious three-dimensional effect.

Harbor depths are shown in fathoms at low water. Gauld recorded that there was generally only one ebb and one flood tide in twenty-four hours; they were considerably affected by the wind, and he observed no regularity in their frequency. The tidal current ran with such great velocity through the mouth of the harbor that the water rose there about three feet; within the harbor, however, this range was only two feet, and scarcely one foot on the bar. The shores of the bay had been washed away in several places. On the west side, exposed as it was to the northeast winds, large tree roots had been found "in the ground upwards of 100 foot within Water-Mark." He noted that the bottom of the harbor and a great part of the anchorage outside the bar along the coast was "deep thick Mud," but the bar was hard sand. As to marine life, Gauld had a serious concern for ships without copper bottoms: "The Harbour and all the Coast abound with Worms." Gauld's "Remarks" conclude with small illustrated keys drawn to indicate swamps, pines and cedars, live oak, hickory, and the observation that "There are many large Magnolia Trees, Sweet Bay, Rhododendron, &c. &c."

A close examination of the "Plan of the Harbour" discloses many interesting features. As the title suggests, it was drafted to demonstrate the maritime utility of this particular body of water, so the drawing does not include Escambia and East bays but delineates only that part of Pensacola Bay south of English (now Emanuel) Point and the entrance from the sea, encompassing those areas which were capable of accommodating the largest vessels whose drafts would permit safe transit of the bar. Rather extensive soundings, along with a projected track line, indicate the deepest water over the bar and between the shoals of the entrance into the bay.

The "Leading Mark over the Bar" is identified as Reid's Tree, almost a mile to the north of the "Red Sandy Cliffs." The name attached to the leading mark honors Robert Reid, master of *Tartar.* Obviously this tree must have been easily identifiable from offshore because of its distinctive characteristics. Gauld drew a straight line from the deepest water over the bar to Reid's Tree. Plotting this from the compass rose near the center of the drawing discloses a bearing of N¼W. Thus a vessel arriving from sea might approach the bar in a safe depth until Reid's Tree was on the indicated bearing and thereupon steer this course toward the leading mark over the deepest water of the bar — three and three-quarters fathoms or twenty-two and one-half feet. Gauld's 1764 drawing is in substantial agreement with the sailing directions later set forth in his "General Description of West Florida, 1769," in which he advised:

In coming from the Eastward or Westward it [is] *best to keep in 6 or 7 fathoms till Reid's Tree is in one* [in line] *with the West declivity of the highest part of the Red Cliff, bearing about N½W as above, and* [then] *steer right in, in that direction. The water Shoals gradually from 4 to 3¾ fathoms; on the shoalest part is 21 feet, then it regularly deepens and the bottom grows softer.*[4]

Gauld noted other prominent features on his manuscript drawing, many of which can presently be identified under different names. He gave ample warning to the seafarer of the bar, breakers, and dry sand, the two latter hazards on the west side of the entrance and close to the projected track line. He delineated "A large Lagoon" and another "Lagoon" now known as Big Lagoon and Bayou Grande, respectively. Near the west end of the "Red Sandy Cliffs" a watering place was indicated, while the point of land on which the wharf of the present Naval Air Station stands received the name of Tartar Point, obviously in honor of Lindsay's ship. Gauld outlined the mouth of present Bayou Chico but left it unnamed. He represented Pensacola with an oblong stockade having bastions at each corner, around which a few houses were placed at random. A comparison of its location with modern sources demonstrates that the village stood in what is now known as the Historical District of Pensacola.[5]

---

4. George Gauld, "A General Description of the Sea Coasts, Harbours, Lakes, Rivers &c. of the Province of West Florida, 1769," American Philosophical Society, Philadelphia, pp. 18–19, cited hereafter as "A General Description of West Florida, 1769."

5. George Gauld, "A Plan of the Harbour of Pensacola . . . 1764"; USC & GS chart 1265,

George Gauld's Pensacola

Leading from the north side of the stockade was the road to Mobile. An "Indian Town," probably the customary camping ground of visiting redskins, was depicted immediately to the east of the stockade, as was Indian Point, some two hundred yards beyond. Along the shoreline, trending to the northeast from this point, a brick kiln, Salt River, and finally English Point were identified; the last two are presently known as Bayou Texar and Emanuel Point. In his "Description of West Florida, 1769," Gauld noted that "The Town of Pensacola is surrounded as it were, by two pretty large runs of water, which take their rise under Gage Hill, a small mount behind the town, and discharge themselves into the Bay, one at each extremity of the Town." Although he depicted them on his drawing, he failed to give names to these streams.[6]

At the entrance of the bay Gauld delineated part of the Island of Santa Rosa and located the signal house one and one-third miles from its western extremity. He outlined the long, narrow peninsula separating the bay from the "Channel of Santa Rosa," and to its promontories, presently known as Deer, Fair, Sand, and Butcherpen points, he gave the names of East Point, Deer Point, Sandy Point, and Oyster Point, respectively. The southern end of the modern Pensacola Bay bridge and the town of Gulf Breeze are located just east of Sand Point.[7]

With George Gauld's detailed survey at hand, Johnstone and his colleagues needed no more than a formal view of Pensacola Bay in order to reach a decision regarding the future of the capital of West Florida. At a meeting on January 30, the governor and council accepted and approved a report on the survey of Pensacola Bay made "in conjunction with Sir John Lindsay." The results of this examination were summarized, signed by Johnstone, Lindsay, and three members of the council, and entered into the record: "We whose names are here unto subscribed, after having taken a Survey of this Bay, are of Opinion that the present situation, every advantage and disadvantage considered, is most Eligible for erecting a town." Having agreed upon the site, the governor and council moved to "take a

---

Pensacola Bay and Approaches; James B. Schaeffer, *Historical District Archaeological Survey* (Historic Pensacola Preservation Board, 1971), pp. 34–36.

6. Gauld, "A General Description of West Florida, 1769," p. 21.

7. Gauld, "A Plan of the Harbour of Pensacola . . . 1764"; USC & GS chart 1265, *Pensacola Bay and Approaches.*

view of the ground surrounding the fort on Friday next in order to give instructions to the Surveyor [Elias Durnford] for the distribution of the streets and lots on the plan he was directed to prepare."[8]

Decision-making continued at a rapid pace when the governor and council, at their meeting on February 3, accepted and approved Durnford's plan. They agreed to reserve specified areas for public buildings and for military and naval purposes and settled on ten provisions under which town lots would be granted. In order to apprise the public of these developments, they authorized the publication of the following advertisement:

These are to give notice that agreeably to a Plan of the town of Pensacola as settled by Mr. Durnford and approved by the Governor and Council, Petitions will be received on Thursday for town lots, from such as may be inclined to take up the same and who may be in a condition to comply with the terms prescribed: which Terms together with the plan are to be seen this day from Eleven of the Clock at the Surveyor General's office. It will be necessary for the Petitioners to specify the number of each Lot which they respectively may petition for.[9]

Only one matter remained to be resolved before the petitions could be acted upon: how to distribute the lots. Apart from those persons having preference by virtue of claims of purchase from previous Spanish owners, and the privileged status enjoyed by officials of the province, other petitioners were divided into classes according to their ability to improve their property and otherwise comply with the provisions set forth in council. Lots were then granted by lottery.[10]

Whatever George Gauld's thoughts of the future, and he must at first have considered himself a temporary denizen on the Gulf Coast, he lost little time in availing himself of the opportunity to become a landowner. At the council meeting of February 5, 1765, the first order of business was action on "a Petition from George Gauld praying for a Grant of that Tract of Land lying to the Westward of a Southeast Line Drawn from Sandy Point to the Channel of Santa Rosa, including all from Sandy Point to the Lagoon." Gauld's peti-

---

8. Council Minutes, January 30, 1765: C.O. 5/625. See Robert Gray, "Elias Durnford" (master's thesis, Auburn University, 1971), p. 19.

9. Council Minutes, February 3, 1765: C.O. 5/625.

10. Ibid., February 4, 7, 9, 1765: C.O. 5/625; Cecil Johnson, *British West Florida 1763–1783* (New Haven, 1943), pp. 29–30.

tion for this tract was granted, "having attention to the reservation which will be necessary for naval purposes & fortifications which the Surveyor will be Instructed to make, Reserving also the Liberty of hawling the sean [seine] to all the Inhabitants of this Province." On May 11, Governor Johnstone ordered the provincial surveyor to lay out a "plantation" of about 720 acres for Gauld, and on February 3, 1766, Durnford certified the completion of the survey and indicated the boundaries of a property totaling 608 acres.[11]

This tract of land lay directly south of and across the bay from Pensacola and, according to the description, comprised the whole western end of the peninsula separating Pensacola Bay and modern Santa Rosa Sound. Bounded by water on three sides, this neck of land obviously was considered of some naval and military value as suggested by the restrictions set out in the grant. Implied in the description of the property was conveyance of title to Gauld of the entire end of the peninsula, but Admiral Sir William Burnaby did in fact reserve a portion of the north shore for the navy's use; just four days after Gauld received his grant the governor in council awarded lots within this area to Joseph Smith, boatswain, and James Crombye, carpenter of H. M. S. *Alarm*, and Thomas Maistell, the gunner of that vessel. These lots fronted on Roebuck Bay on the northern side of the western extremity of the peninsula and were to be used "to build a wharf for the Conveniency of heaving Down [careening] Vessels," according to the language of the petition. In "A General Description of West Florida, 1769," Gauld noted that Admiral Sir William Burnaby, who came to Pensacola as part of his effort to "visit the different stations within the limits of my command, particularly the coast of Florida," drafted plans for such an installation, but nothing ever came of it. In 1769 there was "a careening wharf for small vessels round Deer Point within the entrance of St. Rose's Channel," but it was private property.[12] Ownership of the peninsula westward

---

11. Council Minutes, February 5, 1765: C.O. 5/625; Durnford's warrant, certificate, and plat: C.O. 5/604; Clinton N. Howard, *The British Development of West Florida 1763–1769* (Berkeley, 1947), pp. 35, 55. Gauld's grant, according to its description, comprised the present Town Point area as shown on USC & GS chart 1265 (14th ed., 1968), *Pensacola Bay and Approaches*.

12. Council Minutes, February 9, 1765: C.O. 5/625; Burnaby to Admiralty, June 4, 1764: ADM 1/238; Howard, *British Development*, p. 63. Gauld, "A General Description of West Florida, 1769," p. 24, says of Roebuck Bay: "Just opposite of the town of Pensacola there is a small bay with deep water to the eastward of Deer Point [presently Fair Point]." This is shown on Chart 1265 as Old Navy Cove.

G^{eo.} Gauld — Cartographer

of Gauld's line is clarified by an examination of his charts and Durnford's plat of the property. The *Chart of the Bay and Harbour of Pensacola*, published in 1780, indicates the metes and bounds of several tracts of land, although it fails to ascribe ownership to any, including that which should be Gauld's. Manuscript surveys dating from 1766 and 1768 locate "Gauld's Pl[antatio]n" very precisely half a mile down the shore from Deer Point, and Durnford's drawing shows the property line.[13] It seems unlikely, however, that the naval surveyor ever attempted to develop his Gulf Breeze estate.

Gauld's second piece of real estate, a town lot in Pensacola, was secured less than a week after his first acquisition. He was in the group of "first class" grantees who drew lots on February 8, 1765, and by council action on February 9 received choice town property. To Gauld went lot number 175, 80 × 160 feet, located at the intersection of George and Granby streets at the northwest corner of the open ground outside the fort. With this lot went a larger "garden lot" beyond the northern limits of the town.[14]

It was probably some time before George Gauld undertook to erect his house in Pensacola, but the wisdom of Governor Johnstone's decision to build his capital on the site of Spanish Pensacola was soon confirmed. The vigor with which the first British settlers went to work is suggested by the governor's report to the Board of Trade little more than a year later. Johnstone's natural exuberance and exaggeration are evident, but he could honestly boast a record of achievement and take personal pride in the accelerated tempo of growth in British Pensacola:

It is hardly possible to credit the advancement Pensacola has made in the small Time we have given out Land. There is already one hundred and

---

13. *A Chart of the Bay and Harbour of Pensacola in the Province of West Florida* in J. F. W. Des Barres, *Atlantic Neptune* (London, 1780); "A Survey of the Bay of Pensacola 1766," Library of Congress Map Division; "A Plan of the Bays of Pensacola and Mobile . . . 1768," Hydrographic Dept., Ministry of Defence, Taunton, Eng., D964, Rt. On this last chart Gauld changed the name of East Point to Warburton Point in honor of Lieutenant Charles Warburton of *Sir Edward Hawke*, with whom he had recently sailed. Durnford's plat in C.O. 5/604.

14. Howard, *British Development*, pp. 42, 58; Council Minutes, February 9, 1765: C.O. 5/625. According to Mrs. Lucius V. Ellsworth, Historic Pensacola Preservation Board, letter to John Ware, October 12, 1971, lot No. 175 was on the northwest corner of present Palafox and Intendencia streets. See also "Plan of the Town of Pensacola," copy from the General Land Office, Washington D.C., on file with the Board of Trustees of the Internal Improvement Trust

thirteen good Houses built. Insomuch, that a Water[front] *Lot, without building sells at three hundred Dollars. Such is the Spirit that prevails.*

I am persuaded the Cause of these Improvements is, the Limitations and Restrictions in the Terms of the Grant, which I partly borrowed from the Restrictions after the Fire of London. Their Lordships will observe . . . that the Town Lot must be built on, and the Garden Lots drained, in three years from the Date of the Grant, otherwise there is a large Quit Rent, and the whole is forfeited. So that if a few Examples are made, to shew we are not to be trifled with, the whole void will be filled in five or six years. For men seldom know, till they are pushed, the extent of what they can perform.

By such means the governor claimed to have inspired both his friends and enemies alike. The latter, of whom there were many, were obliged, he boasted, to "build monuments to my Fame." "Their Gardens flourish and their Houses mount as fast as those of my Friends; which they will not deny," he added. Indulging his penchant for allegory and literary allusion, Johnstone misquoted the French philosopher Voltaire whose goodman Candide sagely remarked, *"Mais! Travaillions nous notre jardin."* The governor agreed: "the Person who raises the most Corn and Cabbage for the present is the best man amongst us."[15]

The coastal areas around Pensacola also came in for praise from the governor. Deposits of the finest clay for making bricks had been discovered and also inexhaustible quantities of marl for enriching the kitchen garden plots. The quantity of tar that might be extracted from the pitch pine was alone sufficient to render an estate valuable. Expanding upon the subject Johnstone declared that within twelve miles of Pensacola and around its extensive bay there were at least twenty thousand acres of "as rich land as is upon the face of the earth."

The proud governor was scarcely less expansive in acknowledging the source of his information and evaluation of the region: "It is to Mr. George Gauld, Surveyor on Board His Majesty's Ships, we owe all these Discoveries in the Bay. It is with great Pleasure I acquaint

---

Fund, Tallahassee, Fla.; Elias Durnford's "Plan of the New Town of Pensacola and Country adjacent," P. K. Yonge Library of Florida History, University of Florida, Gainesville.

15. Johnstone to the Board of Trade, April 2, 1766: C.O. 5/574. Candide's remark: "But we must cultivate our garden."

G^{eo.} Gauld — Cartographer

Their Lordships, he is indefatigable in his Profession, and a very worthy subject."[16] If previously the governor and council had failed to credit Gauld for his part in the initial Pensacola surveys, the omission was now corrected; and since much of Johnstone's correspondence, when referring to individuals, reflects the vitriolic outpourings of his quarrelsome nature, it is most significant that George Gauld and his work should be singled out for praise in Johnstone's long report to the Board of Trade. Obviously the governor held his fellow Scot in high esteem. Further evidence of his good opinion is found in a short letter to the board, written little more than a month later:

I now transmit . . . an accurate Plan of the Bay of Pensacola, and a Sketch of the Environs, for the Board of Trade.

This survey has opened a new Light to us, both respecting the Nature of the country, and the Importance of it, which exceeds the most sanguine Expectations.

Perhaps few Maps were ever made with more Exactitude. It is to Mr. George Gauld, Surveyor of the Coast on Board the Fleet, to whom we owe this Survey, who is ever indefatigable in his Profession, and in stolen Hours has done this Public Benefit. He is by far the superior Mathematical Genius among those who have come into the Country. The Surveyor Mr. Durnford, tho' an excellent Draftsman, is much to seek in Mathematics. And besides, he has been so much employed in laying out particular Plots of Land, that I have been unable to procure anything from him on a general Scale.

The Plan of Espiritu Sancto has also been transmitted home to the Admiralty. There requires no further Proof than those two Maps of the Abilities of Mr. Gauld in his Profession. And seeing how rare it is, that Health, Inclination, Mathematical Exactness, and a proper Delineation are found to unite in the same Person, I humbly offer to mention this Gentleman to the Board of Trade, as a Treasure not to be lost; more especially, as his Business in the Fleet will probably be compleated in six months, and that I understand Mr. De Brahm, appointed to survey this Country, is old, infirm, and uncapable of that Fatigue which so great a Work requires.[17]

16. Ibid. Gauld's perseverance is documented by repeated notices in *Active*'s log for January 1766 that her longboat was employed "up the [Escambia] River with the Surveyor on duty." ADM 52/1124.

17. Johnstone to Board of Trade, May 4, 1766: C.O. 5/575. William De Brahm was forty-

George Gauld's Pensacola

Gauld did in fact perform further surveys of Pensacola Bay, its tributaries, and the surrounding countryside, from which he expanded his earlier "Plan of the Harbour of Pensacola" into the more comprehensive and informative draft referred to by the governor as "an accurate Plan of the Bay of Pensacola, and a Sketch of the Environs." Although the governor's version of the title does not agree exactly with Gauld's "A Survey of the Bay of Pensacola 1766," upon which appear two insets entitled "Plan of Pensacola" and "Plan of Campbell-Town founded in Feby. 1766," they are unquestionably the same work. The "Plan of Pensacola" depicts and names some sixteen streets, while that of Campbell Town shows a village laid out in ten blocks by unnamed streets.

On this manuscript drawing Gauld's delineation, if not his ascent of "The River Scambia, or Escambe, called by the Indians Konika," ended some twelve statute miles above its mouth, at which point he noted, "A little farther up the Country abounds with Cane-Swamps. The banks of the Scambia are covered with canes all the way." Some fifteen miles north-northwest of Pensacola he reported that "the soil seems to be good for a considerable extent, being brown Mold mixt with gravel, and producing plenty of grass among the Trees." Gauld's extended examinations of Pensacola Bay and its environs in 1765 and 1766 were incorporated in this "Survey" for the Board of Trade.[18]

As the capital and principal harbor of West Florida, Pensacola received further attention in the cartographic records of Gauld's expeditions along the coast and bays eastward to Cape San Blas in 1766 and that to the west in 1768 which included Mobile Bay. On the first of these, at a distance of some fifteen miles from the mouth of the present Blackwater River near Milton, Florida, Gauld noted: "There are several considerable Rivulets that run into the Oyuva-lana. It is from 2 to 4 fathoms deep in general and about 120 feet broad. The banks of it are swampy and covered with Cypress, Pines, Etc. The water of it is quite fresh where it enters the Bay of Pensacola." Incomplete investigation of the mouth of the present-day Yellow River

---

seven when he was appointed surveyor-general of the Southern District of North America in 1764. Johnstone was unduly pessimistic regarding De Brahm's endurance. See Louis De Vorsey, Jr., *De Brahm's Report of the General Survey in the Southern District of North America* (Columbia, S.C., 1971), pp. 1–6.

18. Copies of "A Survey of the Bay of Pensacola 1766," signed "Geo: Gauld fecit," are found in P.R.O. C.O. 700, Florida No. 32, and the Ministry of Defence Naval Library, Guard Books of MS Maps and Charts, II, No. 47.

apparently resulted in Gauld's confusing and erroneous conclusion that the modern Blackwater and Yellow rivers were one and the same. On his manuscript charts of 1766 and 1768 he depicted only one river flowing into today's Blackwater Bay, and he called it "Uyuva-Lana or Yellow Water"; elsewhere he referred to the same river as the "Middle River."[19]

While Gauld's activity under Sir John Lindsay's command centered upon Pensacola harbor, it is possible that he extended his observations some distance by the end of 1764. On November 6, Lindsay contracted with Edward Mease for the use of the little schooner *Whim* at a fee of £20 per month. From that date until January 11, 1765, Lindsay employed the vessel along the coast. He visited St. Joseph and St. Andrew bays and declared them unsuitable for capital ships. In December he reversed direction, charted Ship Island (the only safe anchorage on the coast west of Pensacola), and made his way to New Orleans, which he found unimpressive. Although there is no specific evidence that Gauld accompanied Lindsay on these excursions, he was a member of *Tartar*'s company and subject to the wishes of a captain who was eager to advance the survey and also to keep his surveyor busy. In any case, these were little more than hasty exploratory ventures; the real work would come later. At the end of January 1765, *Tartar* and Sir John Lindsay sailed from West Florida for careening and repairs at Port Royal. Gauld remained at Pensacola and was transferred to H. M. S. *Alarm,* Captain Rowland Cotton, with whose men and boats he would continue his work.[20]

Naval routine at Pensacola was interrupted on May 10, when two vessels were sighted in the offing over the low-lying sands of Santa Rosa Island. The larger ship, H. M. S. *Active,* flew the distinguishing red pendant of the commander in chief of the Jamaica Station at the mizzen-topmast head, thus heralding the long-expected arrival of

---

19. "A Survey of the Coast of West Florida from Pensacola to Cape Blaise including the Bays of Pensacola, Santa Rosa, St. Andrew, and St. Joseph, with the Shoals lying off Cape Blaise. By George Gauld, M.A. For the Right Honourable the Board of Admiralty. 1766," in MODHD, A9464, 31c. Also "A Plan of the Bays of Pensacola and Mobile with the Sea Coast and Country adjacent. By Geo: Gauld M.A. For the Right Honble. The Board of Admiralty. 1768," in MODHD, D964, Rt. Gauld, "A General Description of West Florida, 1769," p. 23. Compare with USC & GS Chart 1265. Various spellings of the Indian name appear on the several charts.

20. Lindsay to Admiralty, January 30, October 5, 1765: ADM 1/205; lieutenant's log, *Tartar:* ADM L/T/232 (NMM); Muster table, *Alarm:* ADM 36/4951. Because of his subsequent quarrel with Lindsay, Admiral Burnaby refused payment of the £43 due Mease.

Rear Admiral Sir William Burnaby. The flagship, accompanied by the sloop *Ferret*, was greeted by a salute of thirteen guns from *Alarm* as she entered the harbor.

After Admiral Burnaby paid his respects to the governor, he familiarized himself with Pensacola Bay, in company with Captain Cotton of *Alarm*, whom he had summoned, and probably with Gauld, whose chart they would have consulted. Their tour included an examination of the tract of land at Deer Point, previously granted to Gauld with reservations protecting its use for "naval purposes and fortifications." Here Burnaby designated the location of a wharf, but nothing was ever done about it, according to Gauld.[21]

From June 14 until September 4, 1765, Gauld was absent on a surveying trip to the Bay of Espíritu Santo — Tampa Bay — in East Florida.[22] Returning from this assignment, he paused only long enough to draft fair copies of the survey; then he returned to his investigation of the Pensacola area. On December 30 he was transferred to H. M. S. *Active*, Captain Robert Carkett, which served as his base ship until May 4, 1766.[23] During this period numerous references in *Active*'s logbook indicate that her yawl, longboat, or barge had been sent "up the river on duty" with the surveyor, thereby confirming that these estuaries were the last to be explored by Gauld. It is clear that his examination and charting of Pensacola Bay and environs was completed before May 4, 1766, the date Governor Johnstone sent the "Plan of the Bay" to the Board of Trade.[24]

Taken together, these manuscript drawings formed the basis for the only known work by George Gauld published during his lifetime, *A Chart of the Bay and Harbour of Pensacola* in J. F. W. Des Barres's *Atlantic Neptune* in 1780. Gauld's *Chart* demonstrates that he surveyed the lagoons of the north and west shores of lower Pensacola Bay with great care. On the *Chart*, Gauld's earlier designation of "Large Lagoon" became "Grand Lagoon"; his "Lagoon" was christened "Cox's Lagoon"; the unnamed inlet, of which only the mouth was shown earlier, appeared as "West Lagoon"; and his "Salt River" was changed to "East Lagoon." The tributary streams were drawn and shaded in

21. Captain's log, *Alarm:* ADM 51/3757; Gauld, "A General Description of West Florida, 1769," p. 24.
22. Master's log, *Alarm:* ADM 52/1124. See chap. 4 infra.
23. Muster tables of *Tartar* and *Active:* ADM 36/6850 and 36/7543.
24. Master's log, *Active:* ADM 52/1124.

such a manner as to indicate the topography of the land. North and west of Pensacola six lakes were described, further indicating the extent to which Gauld surveyed the surrounding countryside. Modern Escambia, Blackwater, and East bays were examined and charted to their headwaters. The Escambia River was called "Scambia," but the bay itself remained unnamed. From English Point northward, "Gull" and "Stony Points," now known as Devil and Lora points, were identified, and a "Brickkiln" was located just south of the latter promontory. The only other village on the bay, Campbell Town, was identified and located about one mile south of the mouth of the Escambia River. Present Mulatto and Indian bayous were sketched but left unnamed, while the southern extremity of the peninsula, now terminating in Hernandez, Garcon, and White points, was labeled "Yemassee Point." Blackwater Bay retained his Indian and English names of "Oyuva Lana or Yellow River." Unaccountably, Gauld's former "East Point" was changed to "Fan Point" in the published version of his work.

The "Remarks" on Gauld's earlier "Plan" were slightly abbreviated, and the *Chart* was drafted to a smaller scale by Des Barres; coordinates of latitude and longitude were provided in the margins, though both were somewhat in error. Two compass roses were printed, and the magnetic variation was again given as 4°30′ E. Some eighty tracts of land of various sizes and shapes were depicted on Des Barres's version, exclusive of the lots comprising the villages of Pensacola and Campbell Town. Finally, it should be noted that comparisons of Gauld's earlier manuscript drawings and published chart with their modern counterparts disclose amazing similarities as to distances and configurations, a striking demonstration of Gauld's skill as a marine surveyor — and the remarkably few significant changes wrought by man or nature in the intervening period of more than two centuries.

In little more than eighteen months George Gauld, the untried Admiralty coastal surveyor, had proved his worth. He had surveyed lower Pensacola Bay and incorporated the efforts of his immediate superior into the "Plan of the Harbour" on which the decision was made to retain the site of Spanish Pensacola. Although his work was interrupted by a sojourn of more than two and one-half months elsewhere on the coast, Gauld steadily expanded his surveys to include the entire bay, its tributaries and environs. From this he drafted a more comprehensive "Survey" which earned for him and his work the praise of the governor and commanding admiral. Not unlike the legendary phoenix, a new Pensacola was rising — if not from the ashes of the old, then from the bark and palmetto-thatched huts of the earlier Spanish town. George Gauld could readily believe that he was engaged in charting its future.

CHAPTER IV

Espiritu Santo in East Florida, 1765

The arrival of Admiral Sir William Burnaby in the capital of West Florida in May 1765 afforded him the opportunity of personally ordering the survey of Espíritu Santo, or Tampa Bay, and of conferring with the principals involved. He called a temporary halt to Gauld's work around Pensacola, and on May 14 he hoisted the signal for Captain Rowland Cotton of H. M. S. *Alarm* to report aboard *Active*, the flagship. This was the first of several meetings Cotton was to have with his chief. There is no record of the identity of other participants at these meetings or of their conversation, but subsequent events suggest that George Gauld attended at least one of them. On May 17, the admiral informed Cotton that *Alarm* would act as the base vessel for the survey and that Gauld would accompany him to perform those duties.[1]

On May 29 Cotton was summoned to the flagship to receive final instructions for the forthcoming assignment. He was to victual his ship for four months "and proceed to the Bay of Espíritu Santo on the Coast of Florida, and to take a compleat Survey of that Bay, and to examine if it is fit to receive Capital Ships." In the event that his mission should require more time than anticipated, Cotton was to reserve sufficient provisions to return to Pensacola and there to await further orders. Burnaby confirmed that Gauld had his orders to conduct the actual surveys and that he should be allowed his choice of men to assist him. The admiral emphasized that the finished "plan" was *"to be immediately transmitted to me, and not to be sent to the Secretary of the Admiralty."*[2]

The admiral's personal intervention reflected something more than a natural interest in Gulf Coast exploration. Burnaby took much pride but little pleasure in his command over the Jamaica Station, and he was particularly irked that his subordinate, Captain Lindsay, had seized the initiative with regard to surveying activities on the Florida coast. He complained bitterly of "Sir John Lindsay's having

1. Captain's log *Alarm:* ADM 51/3757; Burnaby to Cotton, Pensacola, May 16, 1765: ADM 1/238; Burnaby to Cotton, May 17, 1765: ADM 1/2051.
2. Ibid.

Espíritu Santo in East Florida, 1765

presumed to send home to their Lordships the Plan of Pensacola Bay without having previously sent it to me for my Observations." Lindsay might fancy himself an authority on Florida waters, but his commander in chief in Jamaica did not share that view.

As to the Compliment he [Lindsay] *thinks to pay himself in regard to his being acquainted with the Coast of Florida, I will venture to assert to their Lordships that without the least of his Assistance, they should have had as great an insight into the advantages of that Coast to Great Britain, and as good draughts of all the Harbours &c. which have been taken by Mr. Gauld by my Orders (and not by Sir John Lindsay) except the Bar of Tartar Point.*[3]

Burnaby was further perturbed to discover that, upon sailing for Jamaica, Sir John had instructed the captains at Pensacola to report directly to him while he was at Port Royal, even though the commander in chief himself was at Pensacola. In retort to that slight, Burnaby authorized Captain Cotton, as the senior captain on the coast at the moment, to assume Lindsay's prerogative and fly a commodore's broad pendant until Sir John returned. George Gauld was apparently able to maneuver safely through the shoals of naval protocol. His work was universally held to be of prime importance, and he won unstinting praise from all parties, but he was careful to submit all of his subsequent work, including the survey of Espíritu Santo, directly to the admiral for his examination and transmittal home.[4]

Tampa Bay, generally referred to by the Spanish as Espíritu Santo, lay virtually neglected by its European claimants for almost two centuries. Then, in the closing years of Spain's first occupation of Florida, two surveys of this body of water were made in rapid succession, each primarily concerned with investigating and reporting on timber resources available to the navy of His Catholic Majesty, Ferdinand VI. The first of these examinations was of only twenty-

---

3. Burnaby to Admiralty, November 12, 1765: ADM 1/238. Burnaby was referring to the bar at the entrance to Pensacola harbor.

4. Ibid.; Burnaby to Admiralty, November 16, 1765, February 22, 1766: ADM 1/238; extract from Burnaby to Stephens, September 21, 1765: State Papers 42/65. Further examples of Burnaby's ire may be found in his letters to William Cornwallis, May 29, 31, 1765: *Prince Edward*'s Letterbook, LBK 63 (NMM). The admiral left Pensacola June 3, and Cotton hoisted his broad pendant June 5. Lieut. John Blankett's log, *Alarm:* ADM L/A/59 (NMM). The quarrel between Burnaby and Lindsay was made up after both returned to England. Lindsay to Admiralty, March 3, 1767: ADM 1/2052.

two days' duration and was completed in the closing month of 1756 by Juan Baptista Franco, a draftsman from the naval arsenal or shipyard in Havana. It was but briefly reported by Franco, and he did not attempt to compile a chart of the bay and its rivers. Perhaps for that reason, Lorenzo de Montalba, *comisario ordenador de marina*, the principal naval minister in Cuba, supported a second survey some four months later, and Rear Admiral Frey Blas de Barreda entrusted it to Francisco Maria Celi, pilot of the Royal Fleet. Celi's expedition was only slightly more extended, but it was minutely recorded in his logbook and journal. Furthermore, he drafted a highly ornate chart of the bay which appears to have been the most accurate of its kind to that date.[5]

Despite the favorable accounts of the timber potential submitted by Franco and Celi, and the chart and navigational data provided by the latter, Spain made no attempt to exploit the natural resources of Espíritu Santo during the remaining six years of her possession of Florida. The works of Franco and Celi served little or no practical purpose but survive as valuable historical documents, shedding the first light on Tampa Bay after almost two centuries of indifference and neglect. Eight years later, Espíritu Santo would be subjected to another survey, this time by its new masters, the British.

H.M.S. *Alarm* was a fifth-rate man-of-war of 683 tons, armed with thirty-two guns. Her authorized complement was 180 men, but she set out on this peacetime voyage with only 129 men, exclusive of George Gauld and his survey party. The 22 men who made up this special cadre appear to have been selected from the crew of *Prince Edward*, then stationed in Pensacola harbor. *Alarm*'s customary draft of about sixteen feet permitted easy and safe passage over the twenty-one- to twenty-two-feet depths at the entrance to Pensacola Bay, but the uncharted bars of Espíritu Santo induced Admiral Burnaby, probably at Gauld's suggestion, to engage a small, shallow-draft vessel to ensure that the surveyor and his party were not only able to cross the bar but were also able to proceed to the innermost reaches of the bay itself. For this service the little schooner *Betsey* was chartered, and she fulfilled her duties in a most creditable manner. In reporting to the Admiralty, Burnaby quoted Gauld as having de-

5. Jack D.L. Holmes and John D. Ware, "Juan Baptista Franco and Tampa Bay, 1756," *Tequesta* 28 (1968): 92–95; John D. Ware, trans. and ed., "Tampa Bay in 1757: Francisco Maria Celi's Journal and Logbook," *Florida Historical Quarterly* 50 (1971–72): 158–60.

clared that "he could not have taken the Plan of that Bay in twice the time he did, had it not been for the assistance of the Schooner."[6]

The master of *Alarm*, James Cook, deserves passing comment for his own achievements. He was not only an accomplished shipmaster but possessed considerable skill as a surveyor and mapmaker as well. His talent in both areas was recognized by Admiral Burnaby and offered to the Admiralty as a mitigating circumstance when Cook was later found guilty by a court-martial of charges involving him and Captain Rowland Cotton. Burnaby advised the Admiralty that:

Mr. Cook has been of infinite service in Surveying & taking different plans of the Coast & harbours of West Florida, is a good Draughtsman & a good officer, I therefore hope their Lordships will restore him, & appoint him to some ship that comes out to this station, as he knows the Coast of Florida extreamly well, & will greatly contribute to the completion of the Survey of the Coast.[7]

Cook's cartographic contribution on *Alarm*'s surveying voyage was a partial sketch of the coast of West and East Florida between Pensacola Bay and a point some sixty miles south of the entrance of Espíritu Santo. He depicted the track line of *Alarm* and the soundings and bottom characteristics found en route. The unexplored and unsurveyed segment between Cape San Blas and Port Richey at the mouth of the Pithlachascotee River was left blank by Cook with a notation on the drawing: "Here I wanted to Survey but a Boat could not be spar'd." He also delineated a part of "Spirito Sancto Bay," though with considerable error as to its northern parts, suggesting that he did not accompany Gauld on the surveys of the upper bay.[8] Indeed, Cook's duties as master prevented his doing so.

With Admiral Burnaby's orders in hand, Captain Cotton readied his ship for the surveying cruise. By June 13, 1765 *Alarm* had received her stores, including four live bullocks, at least six casks of ardent spirits, and over five hundred pounds of fresh beef. At mid-

---

6. Muster book, *Alarm:* ADM 36/4951; captain's log, *Alarm:* ADM 51/3757; Burnaby to Admiralty, November 16, 1765: ADM 1/238.

7. Burnaby to Admiralty, February 22, 1766: ADM 1/238. This was not, of course, the famous Pacific Ocean explorer of the same name.

8. "A Draught of Spirito Sancto with the Coast Adjacent 1765 by Is. [James] Cook": Public Record Office, London. *A Draught of West Florida from Cape St. Blaze to the River Ibberville, with Part of the River Mississippi* was published by James Cook in December 1766. This work has two insets: "A Plan of Pensacola Harbour, with the Marks for going in," and "A Draught of Spirito Sancto and Coast Adjacent." MODHD q86 Rc.

morning she shifted from her anchorage near the town of Pensacola to a position between the west end of Santa Rosa Island and Tartar Point to await a favorable wind and tide. The following morning dawned with fair skies and moderate weather, whereupon Cotton gave the order to heave anchor and make sail, with *Betsey* in close pursuit. At noon, after both vessels had cleared Pensacola bar, *Alarm* took the smaller vessel in tow.[9]

Working their way eastward along the coast, the captain and master ordered soundings taken every half hour in order to pass the entrances of Choctawhatchee, St. Andrew, and St. Joseph bays safely. Soundings were continued as they cautiously felt their way past the treacherous and far-reaching shoals of Cape Blaze (San Blas). Because of contrary winds and calms they did not make their landfall, near the site of modern Bayport at the mouth of Weeki Wachee River, until the morning of the third day after passing Cape Blaze.

A southerly course was then set, and a small boat was sent ahead to sound the irregular, rocky bottom and give a signal if danger were encountered. Proceeding for two days in this cautious manner, *Alarm* anchored from sunset until sunrise each day. Before dawn, June 21, James Cook set out in the schooner *Betsey*, accompanied by the longboat, to search out an entrance to the great bay. At noon he returned in the boat, having left the schooner to work her way into the waters of Espíritu Santo and George Gauld to begin his survey.

The same cautious method was employed to determine whether *Alarm* could safely transit the bar and enter the landlocked safety of the bay. A small boat was sent ahead, later joined by Cook in the ship's cutter, to sound and report the depth of water. Two days were spent working *Alarm* slowly over the twenty-four-foot bar, past the north end of the island they named Egmont, and on to a safe anchorage within the bay. At 2:00 P.M., June 24, *Alarm* was moored, "a cable each way. The Body of Egmont Island WSW ½ S, 5 miles, the Hummock on the next Island to Egmont Island [Mullet Key] North 3 miles."

When George Gauld sailed into Tampa Bay in 1765 it was guarded, as it is today, by a series of low coastal islands, two of which lie within the mouth, the others on either side of the entrance. In two centuries, one of these has eroded almost to the point of disappearing, while others have consolidated to make a single larger island.

9. Details of this voyage are taken from the captain's log, *Alarm:* ADM 51/3757; 1st. Lieut. William Johnston's log, and 2nd. Lieut. John Blankett's log: ADM L/A/59 (NMM).

Espiritu Santo in East Florida, 1765

Gauld proceeded first to survey the islands within and immediately adjacent to the entrance, locating them on his rough draft. With these as reference points he was able to sound the three entrance channels, outline their deep-water contours, and place the shoals in proper relation to the channels and islands.[10] Gauld carefully plotted these channels, giving their least depths and sailing directions which consisted of compass courses, distances, and bearings from the previously surveyed islands, all calculated to lead the navigator past the critical shoals and into the safe expanse of the lower bay. Those desiring to proceed farther, Gauld admonished, should do so "according to the Chart."[11]

The southernmost of these channels, presently called Passage Key Inlet, lay "between the North head of Long Island, called Grant's Point, and the small Island Burnaby, 1½ mile to the north of it." Gauld ascribed to this channel a least depth of sixteen feet on its bar. Burnaby Island was depicted in the form of an elongated teardrop one statute mile in length and one-third mile wide at its northern end. Today it remains only as a sandy tidal flat barely awash at high water, supporting a scattering of sea oats and used only by seabirds.[12]

---

10. In producing fair copies of his survey Gauld drafted at least three separate versions, virtually identical in all but nomenclature and the few notes appearing thereon. "A survey of the Bay of Espiritu Santo in East Florida. By Geo: Gauld M.A. 1765" is document x64 Jv.: MODHD. Two other manuscript copies, varying only in minor detail, are preserved in the NLMD, Guard-book series of manuscript maps, II, No. 41, "A Survey of the Bay of Espiritu Santo in East Florida. By Geo: Gauld M.A. in the Year of Our Lord 1765," and No. 42, "A Survey of the Bay of Espiritu Santo in East Florida. By Geo: Gauld M.A. by Order of Sir William Burnaby Rear Admiral of the Red &c. &c. 1765." For information on the NLMD copies John Ware was indebted to Dr. William A. Cumming.

11. George Gauld, *Observations on the Florida Kays, Reef and Gulf* (London, 1796), p. 24. Although published thirty years after Gauld surveyed this region, the *Observations* appear to derive from his field notes.

12. Ibid.; "A Survey of Espiritu Santo," NLMD, No. 42; USC & GS Chart 1257 (6th ed. rev. April 11, 1955), *Tampa Bay and St. Joseph Sound*. Gauld's "Long Island" is Anna Maria Key and "Grant's Point" is now Bean Point, named after a pioneer family on the island. "Grant's Point" was named in honor of James Grant, governor of East Florida. "Burnaby Island," named for Admiral Sir William Burnaby, now Passage Key, has undergone several transformations of shape and size since Gauld's survey, even dividing into two keys at one time. These changes are reflected in the unpublished survey "Tampa Bay," sheet no. 2, surveyed by Lieutenant Commander L. M. Power, U.S.N., 1843: U.S. Coast Survey, "Reconnaissance of *Tampa Bay*, Florida, 1855"; originals in the National Archives, Washington, D.C. (John Ware was indebted to Dewey A. Dye, Jr., Bradenton, Florida, for copies.) Further change and present appearance of Passage Key are also reflected in USC & GS Coast Chart No. 177, *Tampa Bay, Florida*, 1879; and USC & GS Chart 1257 (18th ed., 1973), *Tampa Bay and St. Joseph Sound*.

G^{eo.} Gauld — Cartographer

"The Second Entrance," as characterized by Gauld, "which is between Burnaby and Egmont Islands, is 1½ mile broad." This channel was somewhat wider than the southernmost, but it afforded perhaps one foot more water. As were the channels, it was well marked by soundings and clearly delineated throughout on his manuscript drafts.

"The Third and Northernmost Entrance, which is the principal one, is called by the Spaniards Boca Grande; it is about 3 miles broad, and lies between Egmont Island, and the small islands, to the north-eastward of it, called Mullet Kays." Having thus described the entrance, Gauld noted that a great shoal running west from Egmont about four or five miles and another to the north of it formed the channel. This channel was about one mile wide until it reached the northern point of Egmont where it narrowed to about one-half mile. "The least water in it is 4 fathoms; it deepens pretty regularly, and in the narrowest part you have 17 fathoms; the North point of Egmont Island is steep to, there are 7 fathoms water within 60 yards of it, and 3 fathoms up and down," Gauld wrote as he concluded his description of the deep-water passage through which the main artery of water transporation, Egmont Channel, now runs.

Apparently trusting that the cartographic appearance of the islands flanking the entrance channels would be sufficient, Gauld described only one:

Egmont Island lies North and South, is about 2 miles long, and better than ¼ of a mile broad. The North end is highest, being about 6 or 7 feet above high water mark: a bank much of the same height, and about 40 feet broad, runs on the west side next to the sea, almost the length of the whole island, within which there is a valley covered with bushes of different sorts, and various plants that afford an agreeable verdure, though the soil is hardly any thing but sand and shells. There are a few fresh water swamps, but the water is not good.

The east-west line of the compass rose on Gauld's manuscript chart extended through the center of the longitudinal axis of Egmont Key and was given a value of 27°38′ N latitude. This is only two and one-half minutes of arc and the same number of miles greater than the actual latitude as shown on a modern chart, attesting to the accuracy of Gauld's observations.

His descriptions of the shoals and natural deep-water configurations offshore and adjacent to Egmont Key, as indeed the description of the island itself, are as appropriate today as they were when Gauld penned them more than two centuries ago. There have been few changes either underwater or on the island save those occasioned by the dredged ship channel and the relatively minor topographic differences seen in the several installations which made up the Fort Dade complex, begun in the closing years of the nineteenth century and now in ruins.[13]

A comparison of any one of the three manuscript copies of Gauld's chart with its modern counterpart discloses that the present V-shaped Mullet Key was then three smaller islands separated by two narrow waterways, one near the northern end and one at the southwestern end. The metamorphosis wrought by time and the elements which transformed these three islands into one of its present size and shape may be noted in part by an examination of two of the earliest charts of Tampa Bay published by the United States government, which depict the northern passage as being open in 1879.[14]

Gauld's surveys of the coastal islands of the area took him south to a position just beyond present-day Longboat Pass and north to John's Pass. It will be noted that apart from certain man-made changes there has been considerable natural alteration in the size and shape of these coastal islands, especially to the northward. The shallow passageways have undergone corresponding changes. This phase of his work took Gauld into modern Sarasota Pass and Boca Ciega Bay, where he examined and sketched the many small keys lying therein. Along the shoreline presently occupied by Pasadena and Jungle Estates in Pinellas County he noted "Oyster Bank" on at least two versions of his manuscript chart. Soundings of these estuarine areas and the shoals offshore completed his work here.[15]

At least two weeks were required to make these surveys, after which Gauld and his crew moved into the bay proper with the schooner *Betsey* and smaller boats, leaving *Alarm* at anchor.[16] He sounded and outlined the shoal water adjacent to Pinellas Peninsula

13. Gauld, *Observations*, pp. 24–25. Compare "A Survey of Espiritu Santo . . . 1765," MODHD, x64 Jv and USC & GS Chart 1257 (18th ed., 1973), *Tampa Bay and St. Joseph Sound*.

14. "A Survey of Espiritu Santo . . . 1765," MODHD x64 Jv; U.S. Coast Survey, Reconnaissance of *Tampa Bay*, 1855; USC & GS Coast Chart no. 177, *Tampa Bay, Florida*, 1879.

15. "A Survey of Espiritu Santo . . . 1765," MODHD x64 Jv, and USC & GS Chart 1257, *Tampa Bay and St. Joseph Sound* (6th ed. rev., 1955); USC & GS, 1256 (5th ed., 1965), *Lemon Bay to Passage Key Inlet*.

16. Captain's log, *Alarm*: ADM 51/3757. An entry of July 5, 1765, two weeks after arrival,

and the great middle ground in the bay to its eastward, noting the fathoms in black figures and the feet in red at mean low water. Delineating the eastern shoreline of the peninsula, Gauld inscribed on two of the fair copies of his manuscript chart, in the area that is now downtown St. Petersburg, the prophetic notation "A pretty good place for a Settlement." Nearby he depicted a small circular pond with the inscription "Fresh Water." This may well have been Mirror Lake.[17] Old Tampa Bay, referred to by Gauld as "Tampa Bay," was quite faithfully rendered as to configuration, size, and depths of water, judging by present criteria.

Near the shoreline, less than one mile west of present Gadsden Point, Gauld depicted and identified a "Fresh Water Pond," and near Ballast Point he noted "Black Rocks" outward from the beach. He sketched three unnamed islands lying off the mouth of present Hillsborough River, as did Celi in his earlier chart; but a map of 1830, reputed to be the first made by Americans in South Florida, shows only two such islands, as does the unquestionably accurate U.S. Coastal Survey chart of 1879. From whatever cause, one of the three islands depicted both by Gauld and Celi disappeared in an interval of sixty-five years.[18] The eastern shoreline of what later was known as Hooker's Point was indicated by Gauld's chart as "Rocky Ground." His version of the promontory compares closely with that on the earlier maps, but all this shoreline has undergone radical change in relatively modern times with the dredging of Davis and Seddon islands and more recent port developments. Gauld's delineation of Hillsborough Bay and the eastern side of Tampa Bay, with its adjacent shoals, indicates careful examination of these waters; the complete absence of any of the several streams which enter these bodies suggests, however, that no effort was expended investigating Hillsborough, Six-Mile Creek, Alafia, Little Manatee, and Manatee rivers.[19] Apart from the differences previously mentioned, comparison of Gauld's cartographic record of Espíritu Santo in 1765 with the U.S. Coast and Geodetic Survey chart of Tampa Bay, 1879, drafted be-

---

notes that a gun was fired and a signal made to the schooner, indicating that *Betsey* was still in the area.

17. "A Survey of Espiritu Santo . . . 1765," NLMD, Nos. 41 and 42, and USC & GS Chart 1257, *Tampa Bay and St. Joseph Sound*.

18. "A Survey of Espiritu Santo . . . 1765"; Karl H. Grismer, *Tampa* (St. Petersburg, 1960), p. 57; USC & GS Chart No. 177, *Tampa Bay, Florida*, 1879.

19. "A Survey of Espiritu Santo . . . 1765"; USC & GS Chart 1257, *Tampa Bay and St. Joseph Sound*.

Espiritu Santo in East Florida, 1765

fore any significant man-made changes had occurred, discloses a striking similarity of configuration, size, and soundings and attests to Gauld's skill as a surveyor and the care he exercised in examining this body of water.

Considering his relatively short stay in Tampa Bay, Gauld's observations of the natural phenomena also compare favorably with those of modern authorities. The arrows on his manuscript chart indicating the direction of the current at flood tide are generally quite accurate, and he correctly noted that:

there are so many Inlets or Openings into the Bay, the course of it is considerably affected thereby. Within Egmont Island the Flood sets very strong to the North East, and continues nearly in the same direction, till it divides itself into the Bay of Tampa [Old Tampa Bay] *and Hillsborough Bay. Without Egmont Island, inshore it runs to the Southward, but in the offing, to the Northward, where it has very little strength.*

Had Gauld remained in these positions long enough he would have observed that these diametrically opposed currents joined to form a clockwise rotary current of considerable daily inequality. His ascription of a velocity of about four knots on the flood "through the Main Channel at the North End of Egmont Island" is, however, at considerable variance with modern tabulations, which seldom record more than two knots at flood tide in this position.

"At Common Tides the Flood rises near 3 feet in the Bay, and about 2 feet a considerable way out in the Channel, but where the Stream is confined, it rises near a Fathom," Gauld reported in describing the range of tides. As to their unconventional nature he correctly observed:

The Tides are very irregular here, and seem to be very little governed by the Moon. Sometimes there are two Tides of Flood and two of Ebb in the Space of 24 Hours, and sometimes more; but most commonly there is only one of each: the Time is quite uncertain and not at all to be depended on, for instance it often happens that one Tide will run 15 or 16 Hours, and the other only 6 or 7 Hours.[20]

20. "A Survey of Espiritu Santo . . . 1765," NLMD, No. 42; *Tidal Current Tables, 1973,* Atlantic Coast of North America (U.S. Department of Commerce, National Ocean Survey, 1973), pp. 112–17 and passim; *United States Coast Pilot 5, Atlantic Coast, Gulf of Mexico, Puerto Rico, and Virgin Islands,* Coast and Geodetic Survey (Washington, D.C., 1967), p. 78.

G$^{eo.}$ Gauld — Cartographer

Gauld's surveys kept him in or near Tampa Bay from June 22 until August 30. He was therefore unable to observe the wind and weather patterns, nor did he encounter any of the violent northwesters which sweep across the Gulf of Mexico, striking the west coast of the Florida peninsula with varying degrees of intensity several times during the winter and early spring months. For the period of his sojourn he noted that the winds were variable, generally on the "Southern Board," as he termed it, seldom passing northward of northeast and northwest. The usual "moderate breezes" of these months were often attended by rain and thunder squalls but were rarely interrupted by gales. "The Thermometer was generally from 82 to 83 [degrees] at a medium." The weather has changed little since the eighteenth century.[21]

Gauld noted an almost complete absence of fresh water near the mouth of the bay, "but by digging and sinking casks" water might be obtained, though sometimes it was a little brackish after high tides because of the low elevation of the islands. Two versions of his manuscript chart indicate that *Alarm* secured her water about one mile from the northeast end of present Mullet Key. "Egmont Island is considerably higher than any of the rest," Gauld added, "and affords pretty good Water by digging." Almost as an afterthought he advised that "A small Fort on the North End of this Island could easily command the Entrance of the Harbour." More than a century later the third Fort Dade was constructed on Egmont Key, as was its counterpart, Fort DeSoto on Mullet Key, neither of which ever fired a shot at an enemy.[22]

Gauld wrote but briefly of the topography and flora of the area. He recorded that mangroves covered most of the islands, several of which "overflowed at high Tides." He further noted that the east side of the bay was likewise very low and abounded with mangrove swamps until they gave way to higher land and a profusion of pine trees. Differing in some degree was the land on the west side of the bay and around its two main arms, present Old Tampa and Hillsborough bays. In general, the elevation was higher a short distance from shore and the land was "covered with Pines and some Live

21. "A Survey of Espiritu Santo . . . 1765"; José Carlos Millas, *Hurricanes of the Caribbean and Adjacent Regions, 1492–1800* (Miami, 1968), pp. 18–19; *United States Pilot 5*, p. 261.
22. "A Survey of Espiritu Santo . . . 1765"; Grismer, *Tampa*, p. 210; John D. Ware, "A View of Celi's Journal of Surveys and Chart of 1757," *Florida Historical Quarterly*, 47 (1968): 24.

Espiritu Santo in East Florida, 1765

Oaks." Gauld's rather vague comment—"Here and there are some small Runs of fresh Water"—confirms the view that he undertook no surveys of the streams he encountered along the way.[23]

In bringing his observations to a close, Gauld touched on the fauna of the area and in so doing gave perhaps the first account of civilized habitation on Tampa Bay since Pedro Menéndez de Avilés established at Tocabaga a small garrison which was slain by the natives whom they hoped to instruct in the Christian faith. Gauld wrote:

Espíritu Santo will admit large vessels, and there they will find abundance of fish, oysters, and clams, as well as large and small water-fowls, turkeys, deer, &c. with plenty of fresh water and wood. The Spaniards resort to the Mullet Kays for the purpose of fishing, and have built huts on the principal of them, where there are likewise wells of fresh water. The chief growth on these Kays, and on the large ones at the entrance of the Bay, are mangroves, and blackwood bushes.

Gauld's final comment was in the nature of an admonition to seafarers who might attempt a landfall: "We must observe that the land about the Bay is low, and visible only at 8 or 9 miles distance, where you will have from 7½ to 7 fathoms water."[24]

James Cook contributed in a minor way to the sum total of information obtained on this voyage of exploration when he noted on his manuscript drawing, "All the coast in this Draught is a low Sandy Barren Soil. Here is great plenty Deer & wild Cattle, all sorts of Game & Fish. We found Indian Hutts but no person in them." Interestingly, eight years earlier, Celi had indicated the likelihood of wild cattle in the area by depicting a bovine animal on his highly ornamental chart near modern Bradenton.[25]

And what incidents befell *Alarm* while she remained at anchor for sixty-eight days in lower Tampa Bay? Fourteen of her complement died of undisclosed causes—ten marines and four seamen. On the final day, August 29, a smoldering quarrel between the captain and the master blazed up, and James Cook was confined to his cabin by

23. "A Survey of Espiritu Santo . . . 1765."
24. Gauld, *Observations*, p. 25; Woodbury Lowery, *The Spanish Settlements within the Present Limits of the United States* (New York, 1905), pp. 280, 342, 448–50.
25. Cook, "A Draught of Spirito Sancto with the Coast Adjacent, 1765"; Francisco Maria Celi, "Plano de la Gran Bahia de Tampa, Nuevamente Sn. Fernando . . . 1757," Museo Naval, Madrid, Spain; Ware, "A View of Celi's Journal," p. 23.

G^{eo.} Gauld — Cartographer

Captain Cotton's orders. Two men received a dozen lashes each for neglect of duty and drunkenness. A third crewman, Martinico Gomez, suffered two dozen lashes for theft and drunkenness. Perhaps he broached a liquor cask, thus earning for himself the additional punishment for theft.[26]

As mentioned, good weather prevailed throughout *Alarm*'s stay. Ten days after her arrival casks were sent ashore to the innermost of the Mullet Kays "to sink for a well." The crew was employed in various routine duties: overhauling the rigging and sails, clearing hawse, "working up junk," and transporting water from ashore. A sail was sighted in the offing shortly after dawn, August 14, and proceeding without hesitation into the bay, the vessel, a schooner bound from Virginia to Pensacola, came to anchor, probably for the purpose of replenishing her supply of water and firewood. Her arrival suggests that Espíritu Santo and its resources were not unfamiliar to individual shipmasters in the merchant trade.

By August 29 Gauld had completed his surveys, and little *Betsey* had been brought alongside *Alarm* and both sides "hove out" for examination and repairs. Before dawn the following morning, *Alarm* raised her anchor and fired a gun for the smaller vessel to accompany her toward the bar. By noon, *Betsey* had been taken in tow, and both vessels were three to four leagues westward of Egmont Island.

The passage to Pensacola was made without further incident, save when the schooner broke her tow in an afternoon squall, and midmorning of September 4 saw *Alarm* anchored inside the bar awaiting a favorable tidal current and wind. Captain Cotton struck his broad pendant as soon as he observed that Sir John Lindsay had returned

26. Cotton had confined Cook to quarters in April, when they were at Pensacola, for disrespect toward the ship's officers, but they had patched up their differences before sailing. In July, Cook was assigned to investigate the coast north and south of Tampa Bay; he alienated the captain by failing to produce the required report, by returning to the ship each night, and by acting in a mutinous and threatening manner. Both Cook and Cotton brought charges against one another, and courts-martial were held at Port Royal on January 27 and February 5, 1766. Cook defended himself by arguing (truthfully — he was hospitalized when they reached Jamaica) that he was ill with fever much of the time they were in Tampa Bay. The master accused Cotton of having used ship's stores for his private profit and of having falsified the ship's log. Although the evidence in Cook's favor was quite strong, the captains sitting in judgment accepted Cotton's version of the affair, and Cook was dismissed as master and rendered incapable of ever serving as an officer in the navy. The admiral's plea in his behalf seems to have been ignored. Captain's log, *Alarm*: ADM 51/3757; lieutenants' logs, *Alarm*: ADM L/A/59 (NMM); *Prince Edward*'s Letterbook: LBK 63 (NMM); Court-martial records: ADM 1/5303, Pt. 2.

Espiritu Santo in East Florida, 1765

from Jamaica, and *Tartar* was now at anchor in the harbor along with H.M.S. *Prince Edward* and "ten sail of transports."[27]

So ended the first of George Gauld's several special surveys. Three fair copies drafted from his field manuscript graphically record the results of the survey of Espíritu Santo. They are identical in configuration and soundings, but each differs from the others either in place-names, notations, or length of the "Remarks." Common to all is the identification of Egmont Island. Significantly, one of the two otherwise nearly identical copies indicates in the title that the survey was "By Order of Sir William Burnaby, Rear Admiral of the Red &c. &c. 1765" and identifies Passage Key as "Burnaby Island." The dedication on its counterpart has been clumsily altered, obviously not by Gauld's hand, to read "In the year of our Lord 1765." It appears that on this particular copy an effort was made to eliminate all reference to Burnaby, even to the naming of an island in his honor.

Of the eponymous nomenclature appearing on the most informative copy of Gauld's manuscript charts, three names remain in use to this day: Egmont, Hillsborough, and Tampa. Additionally, the single V-shaped island now known as Mullet Key was referred to in a later written description by Gauld as Mullet Kays. Egmont Island was named for John Perceval, second Earl of Egmont, who had been first Lord of the Admiralty since 1763; and at the time of Gauld's survey of the bay, Wills Hill, Earl of Hillsborough, was president of the Board of Trade, thus meriting the honor of having an arm of the bay named for him. Tampa, the oldest of the place-names, is of Indian origin and was the name of a large town of Carlos, chief of the warlike aboriginal Calusa tribe. Although significance has been attached to the word, its meaning is not known.

Gauld was strangely guarded in reporting on the stated object of the survey, that is, to determine whether the bay was "fit to receive Capital Ships." He had sounded a minimum depth of twenty-four feet on the bar and even deeper in parts of the bay itself. Most capital ships, whose drafts were in the twenty-two-foot range, could have passed over this bar, yet Gauld did not make such a recommendation, perhaps because he considered two feet an insufficient margin for safety. He limited himself to the comment that "Spiritu Santo Bay will admit large vessels," leaving the question of naval utility to the wisdom of the Admiralty.[28]

27. Captain's log, *Alarm:* ADM 51/3757.
28. Gauld, *Observations*, pp. 23-25; *Memoir of Do. d'Escalente Fontaneda*, trans. Buckingham

G^{eo.} Gauld — Cartographer

Judging from the known versions of Gauld's manuscript chart, his survey of Espíritu Santo and its branches was thorough and accurate, although the brevity of his written descriptions leaves something to be desired with regard to its natural resources. Governor George Johnstone was quick to recognize the potential of Tampa Bay, and within a fortnight of Gauld's return he had secured "a copy of the survey of Espíritu Santo harbour which I regard as one of the most material advantages of the Peace and one of the most fortunate circumstances for Great Britain." Johnstone considered Tampa Bay "in every respect superior to Havannah," and he entrusted the hastily copied chart to Lieutenant Colonel David Wedderburn for speedy delivery to their great patron the Earl of Bute.[29] Yet Gauld's chart, like much of his subsequent work, lay unused in the Admiralty files, not because of any shortcomings on his part but because of geographic, political, and economic considerations. Magnificent harbor that it was, Espíritu Santo lay far from colonial centers of population, and although the great westward movement was even then manifesting itself on the east bank of the Mississippi River, it would be another half century and more before Tampa Bay and its environs would be opened and safe for permanent settlement. This would occur under a proprietorship then twice removed from the British: that of the burgeoning and acquisitive United States.

Smith (Miami, 1944), p. 30; J. Clarence Simpson, *A Provisional Gazetteer of Florida Place-names of Indian Origin*, ed. Mark F. Boyd (Tallahassee, 1956), pp. 106–9.

29. Johnstone to Bute, September 18, 1765: Bute Correspondence, MS 3.615A (bundle 3), letter No. 145, Central Library, The Hayes, Cardiff, Glamorgan. Robert Rea is indebted to Professor Robin F. A. Fabel for this information. On Wedderburn see Robert R. Rea, "Outpost of Empire," *Alabama Review* 7 (1954): 217–32.

CHAPTER V

Coastal Surveying, 1765-1766

Coastal surveying in the Gulf of Mexico was a seasonal occupation. Uncertain weather from October onward could turn viciously foul, and during the winter of 1765-66, George Gauld might have wished to confine himself to the quiet waters of Pensacola Bay, avoiding the hazards of the open sea. He was usually little affected by the frequent reassignment of the ships operating on the Gulf Coast, although that process required his transfer from one to another; but in September 1765, Sir John Lindsay returned to Pensacola, *Alarm* was sent back to Jamaica, and on October 3 the surveyor was returned to *Tartar*'s muster book.

Determined to advance the Gulf Coast survey in spite of Admiral Burnaby's jealousy and petty intervention, Sir John intended to explore Apalachicola Bay and to take George Gauld with him so that the surveyor should not be idle but should earn his ten shillings a day. Lindsay hired a schooner for £25 a month from John Doyle of Pensacola, and on October 26, the little *Tickel Bender* (the name implies that she was suited for maneuvering in narrow waters) sailed with Captain Lindsay aboard. Whether the naval surveyor accompanied him is not certain. After sounding the shoals off Cape Blaze, Lindsay made for St. George Island and observed that the bar at the mouth of the Apalachicola River was too shallow for safe passage by large vessels. He then crossed Appalachee Bay and turned south to Espíritu Santo. There Lindsay took the latitude and found himself eleven miles off Gauld's calculations, a discrepancy that the surveyor diplomatically attributed to some fault of his declination tables, thereby satisfying his captain who commented upon Gauld's "great exactness" in all other matters. Lindsay was delighted to find that Tampa Bay "answers all our expectations," and he urged that the use of Egmont Island be reserved to the navy. From Espíritu Santo, *Tickel Bender* wandered south among the keys (Lindsay had no idea just where, perhaps Key West) before foul weather forced her to run for home. She reached Pensacola on November 29 and was discharged from the surveying service on December 11, having cost the Admiralty only £31.16.8.[1]

1. Lindsay to Admiralty, September 16, October 25, December 28, 1765: ADM 1/2051.

G^(eo.) Gauld — Cartographer

While Lindsay sailed east and south, Captain Robert Carkett in H. M. S. *Active* was investigating the coast west of Pensacola. Skirting the bar of Mobile Bay, Carkett worked his vessel into the Ship Island anchorage, November 14, and spent three weeks sounding the waters of Ship and Cat islands before returning to Pensacola on December 12.[2]

If Gauld did not accompany Lindsay on his foray, he might well have sailed with Carkett — or remained in Pensacola: the record is unclear. In any case, the coastal survey had been considerably advanced by the end of the year, and Gauld must have welcomed a respite from the sea. Just before Christmas, Lindsay and *Tartar* sailed for home; *Active* became Gauld's ship for purposes of muster and pay from December 30 until May 5, 1766, when Carkett left Pensacola. Captain George Murray's *Ferret* was designated to carry on the survey and anchored at Pensacola on March 7. Gauld seems to have begun working with Murray in April, but it was May 6 before he joined *Ferret*'s company for a six months' tour of duty.[3]

The previous year's experience surveying Tampa Bay had demonstrated that Gauld's work could best be accomplished in a vessel smaller than a navy sloop. He had pursued that point with Admiral Burnaby, and the commander in chief had suggested to the Admiralty that two small schooners be built for service on the rivers and coasts of West Florida. Burnaby recommended vessels of a hundred tons' burden, drawing ten feet of water, manned by six or eight men under a midshipman and a lieutenant, and armed with ten four-pound guns. If built in American shipyards, either in New York or New England, they might be secured for £400 each — cheaper than English construction and cheaper than hiring merchant vessels on a short-term basis.[4]

Neither admiral nor surveyor assumed that the suggestion would be acted upon immediately, and in order that Gauld's work might be

---

The voyage was not unproductive; "A Sketch of the Tortugas & Part of the Martyres taken by Sir John Lindsay" is listed in the manuscript "Catalogue of Drawings & Engraved Maps, Charts & Plans; the Property of Mr. Thomas Jefferys; Geographer to the King, 1775." Copy in the British Museum Map Room.

2. Journal of Lieut. James Sutherland, *Active:* ADM L/A/16 (NMM).

3. Muster tables, *Tartar, Active,* and *Ferret:* ADM 36/6850, 7543, 7376; Carkett to Admiralty, December 30, 1765, April 1, 1766: ADM 1/1609, Pt. 3. According to Carkett, Gauld was discharged into *Ferret* by April 1, 1766.

4. Burnaby to Admiralty, September 21, 1765 (extract): SP 42/65; Burnaby to Admiralty, December 16, 1765, February 22, 1766: ADM 1/238.

Coastal Surveying, 1765–1766

renewed in the spring, a charter craft was once more employed in 1766. For undisclosed reasons, *Betsey* was not retained, even though she had proved quite satisfactory on the survey of Espíritu Santo. In her place Captain Carkett was authorized to purchase the schooner *Charlotta* for Gauld's use. On April 12 and 14, Robert Christian, master of *Active*, went aboard to check her condition. Thereafter she was ballasted, victualed for eight days, and placed under the command of a navy lieutenant. Her company was made up of a midshipman, five seamen, a corporal, and four marines, and her armament boasted two swivel guns, four muskets, thirteen pistols, and two cutlasses.[5]

A shakedown cruise to Mobile in mid-April revealed numerous faults in *Charlotta* which required correction before she would be fit for an extended period at sea. The carpenters of *Ferret* worked the better part of a month on the schooner. They were followed by the sailmaker, who provided new tarpaulins, and by the cooper, who repaired the water casks. *Ferret*'s cutter, also newly fitted, was assigned to *Charlotta* for Gauld's use on the survey. Finally, with the aid of a special working party assembled under some pressure to bring the schooner to readiness, Gauld was able to set off on June 15, 1766, *Ferret*'s log noting: "A.M. Sailed hence the Charlotta Schooner on the Surveying Service."[6]

Gauld's target area was the long, crescent-shaped coastline of West Florida from Pensacola to Cape San Blas. Recording soundings and bottom characteristics for some miles off shore, he then proceeded to delineate the shoals and bars at the entrances to three bays: Rose (now known as Choctawhatchee), St. Andrew, and St. Joseph. The bays themselves were surveyed, and pertinent information relating to each was noted on his composite manuscript drawing of the entire area. Gauld's base ship, *Ferret*, remained at anchor in Pensacola harbor during the two months he was absent on this survey; consequently day-by-day descriptions of his activities are lacking. Captain Murray did, however, secure one indirect confirmation of the surveyor's progress: *Ferret*'s log for July 27 notes, "A.M. Apprehended and Brought on Board Samuel Roan, Seaman, who had Deserted from the Surveying Schooner in Sta. Rosa bay."[7] One must wonder

---

5. Admiral William Parry to Admiralty, March 21, 1767: ADM 1/238; Master's log, *Active*: ADM 52/1124; Journal of Lieut. James Sutherland, *Active*: ADM L/A/16 (NMM).

6. Master's log, *Ferret*: ADM 52/1232; lieutenant's log, *Ferret*: ADM L/F/70A (NMM); Murray to Admiralty, March 31, June 16, 1766: ADM 1/2116.

7. Master's log, *Ferret*.

doubt intended to be recognized as specific vessels to which Gauld was attached. (Courtesy of the Historic Pensacola Preservation Board.)

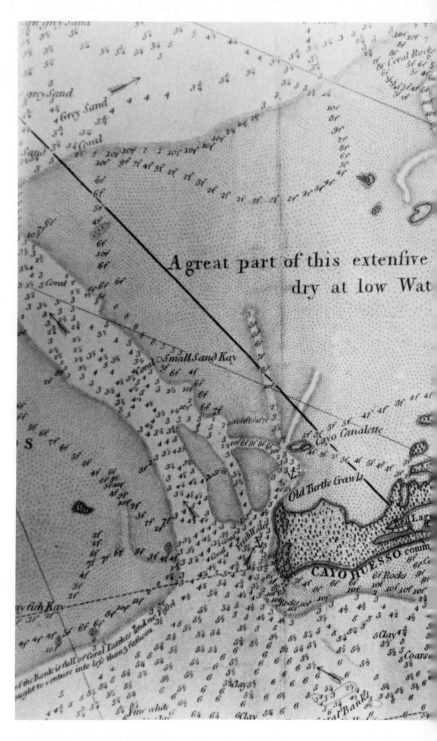

II & III. Key West and the Dry Tortugas. These are small portions of *An Accurate Chart of the Tortugas and Florida Kays* published by Faden in 1790

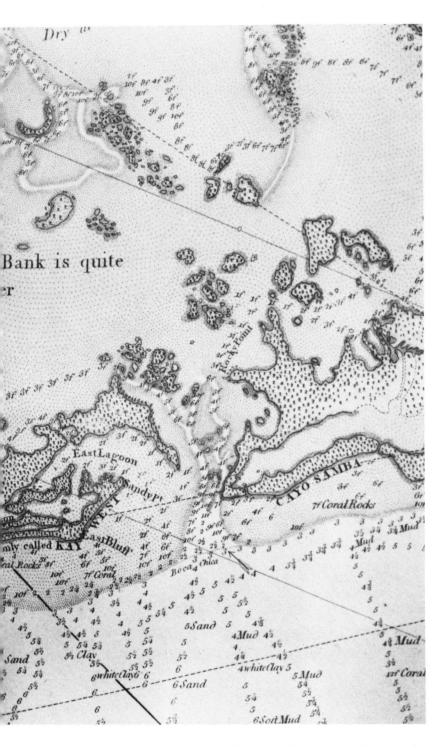

and showing the painstaking work in which Gauld was engaged during the years 1773–75. (Courtesy of the P. K. Yonge Library of Florida History.)

I. "A View of Pensacola in West Florida." In this remarkable panorama of British Pensacola, the surveyor let his considerable artistic talents prevail, but his professional standards guarantee its authenticity. The ships in the foreground were no

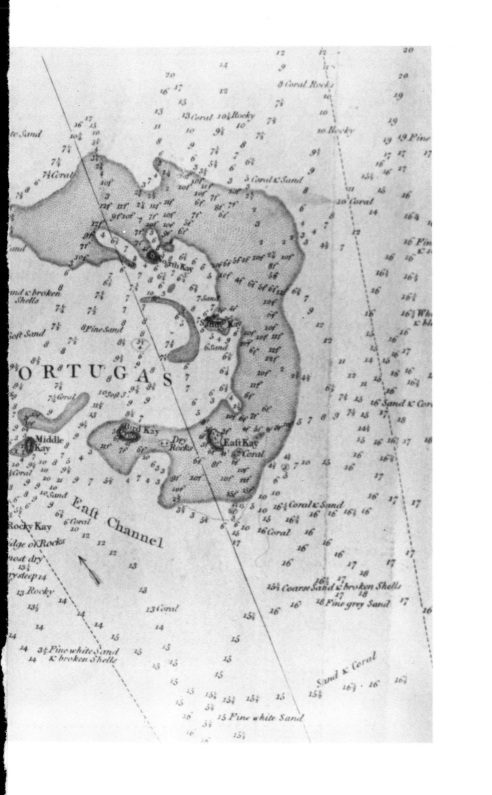

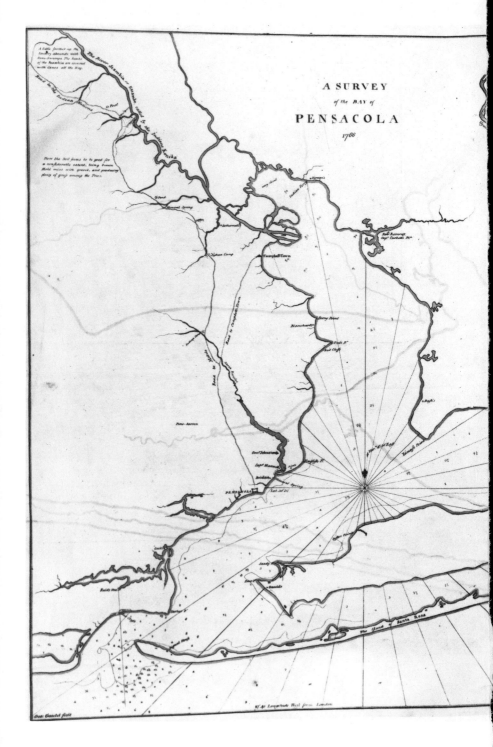

IV. "A Survey of the Bay of Pensacola, 1766." This early chart displays Gauld's precision and attention to details of concern to navigators: the signal house, Reid's tree, the Red Cliffs, the fort (and incidentally Gauld's plantation). The street plans of Pen-

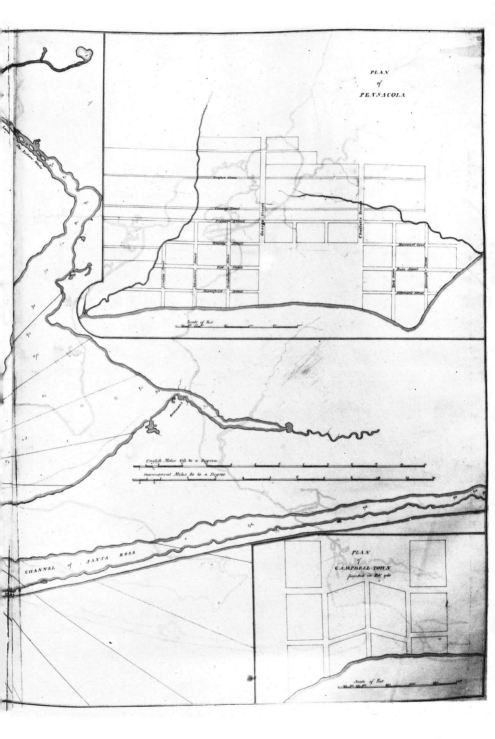

sacola and Campbell Town were drawn up by Elias Durnford. (Courtesy of the Public Record Office.)

III.

about the conditions faced by the survey party that led Seaman Roan to hike back to Pensacola!

George Gauld found the scene quite sufficiently interesting to keep him at work. "The Island of St. Rose is nearly of the figure of an Italian S, the West end inclining a little to the Northward and stretching across the Mouth of Pensacola Harbour, as the East end which turns a little to the Southward extends across the Entrance of St. Rose's Bay." With this figure Gauld began his description of the island which was the first segment of hundreds of miles of coastline he would eventually survey and chart. He estimated it to be about fifty miles long with a width not exceeding one-half mile. It was "very remarkable for its white Sandy Hummocks and straggling Trees here and there." Four tall trees, eighteen miles from the west end, were clumped so close together they appeared as one at a distance. A similar stand was sighted a league to the eastward, and there were other hummocks that more readily lent themselves to observation than to description. "An attentive person after once or twice sailing along can be at no loss to know what part of the Coast he falls in with," Gauld rather optimistically assured the curious navigator.[8]

"The peculiarity of the appearance of Rose Island from the Sea," he continued, shifting his emphasis from the apparent to the unseen, "and the deep soundings all along it are of great Service to know the Coast; there is 9 or 10 fathoms in some places within a mile or two of the Shore; and when a Frigate is in 16 or 17 fathoms the tops of the Trees on the Main Land may be discerned from the Quarter Deck." As for the all-important bottom characteristics, they were generally fine white sand with broken shells and black specks. An exception was found in a considerable area off the east end of Santa Rosa Island, where, out of sight of land, in twenty to forty fathoms, the bottom was coarse gravel mixed with coral. Gauld stressed the importance of this phenomenon as it marked the only spot on the coast having such identifying characteristics. "There is a coral Bottom off the Bay of Espíritu Santo, and some other parts on the [west] Coast of East Florida, but it generally begins in 7 or 8 fathoms within sight of Land, from which, and the Difference of Latitude the one cannot be mistaken for the other."[9]

8. George Gauld, "A General Description of West Florida, 1769," p. 24.
9. Ibid.; and "A Survey of the Coast of West Florida from Pensacola to Cape Blaise: including the Bays of Pensacola, Santa Rosa, St. Andrew, and St. Joseph, with the Shoals lying off Cape Blaise. By George Gauld, M.A. For the Right Honorable the Board of Admiralty,

Coastal Surveying, 1765–1766

Gauld characterized St. Rose or Choctawhatchee as "a very extensive Bay stretching about 30 miles to the NE" with a breadth of four to six miles. The bar before its entrance afforded only seven or eight feet, but a narrow channel deepened to sixteen or seventeen feet as far as the "Red Bluff on the Main Land." Farther in Gauld noted a large shoal, dry in places and yielding only four or five feet at best, blocking the bay side of the entrance. He advised that "nothing but very small Vessels can enter this Bay from the Sea, and the Channel between Rose Island and the Main is just sufficient for Boats or Petty Augres from Pensacola." Despite its limitations, he considered the inside route—Santa Rosa Sound—as the safest communication between Pensacola and Choctawhatchee bays. Near its eastern terminus, at present Fort Walton Beach, he noted on his chart the presence of cedar and live oak.[10] Gauld was probably denied access to the bay in *Charlotta* because of the shoals and used the cutter to conduct his examinations.

Near the present town of Valparaiso, Gauld noted an Indian habitation on his chart, and with it he provided some historical information concerning the natives:

The Coussatas [Koasati], *a small adopted Tribe of the Alabama Indians are settled lately on this part of Sta. Rosa Bay, having left their own Country, on account of the present War between the Chactaws and the Upper Creeks; being unwilling to take part with either. They consist of about 200 Souls, including men, women and Children.*

Other "Indian Camps," perhaps of the same tribe, were indicated on his chart, one near the present town of Shalimar, another on the south side of the bay on Fourmile Point.[11] He later reported Indians living along the banks of the "Chacta-hatchi or Pea Creek" but at some distance from the bay. This river extended to the northeast and

---

1766." Bottom characteristics have been determined since the sounding lead came into use by filling its concave bottom with tallow, soap, or some other sticky material not immediately water-soluble. When the lead strikes the bottom, particles are imbedded in the soft substance, so that not only is the depth of the water fathomed, but retrieval of the lead provides a visual determination of the nature of the bottom.

10. Ibid. The "Red Bluff on the Main Land" is at the eastern end of the Santa Rosa Island–Destin bridge. See USC & GS Chart 1264 (5th ed., 1968), *Choctawhatchee Bay and Approaches*.

11. Gauld, "A Survey of the Coast of West Florida from Pensacola to Cape Blaise . . . 1766"; USC & GS Chart 1264, *Choctawhatchee Bay and Approaches*; John B. Swanton, *The Indians of the Southeastern United States* (Washington, D.C., 1946), pp. 125, 146.

entered "the Bottom of the Bay through Several Mouths, but so Shoal that only a small Boat or Canoe can enter." Of the Indians he wrote:

About 12 or 14 Leagues up this River are settled a small party of the Coussa [Coosa] *Indians who have join'd the Creeks and sometimes bring Provisions and Wild Fowl to Pensacola in their Canoes; this they agree to do by Treaty; but in the Main they are a very lazy sort of People.*[12]

Juniper Creek, emptying into Boggy Bayou at the northwest end of Choctawhatchee Bay, was recorded as being only twenty-five feet broad and canopied by the overhanging branches of trees on either side. Cedars abounded along this swift stream, whose depth averaged about eight feet except at the mouth of the bayou, where it shoaled to two feet. Gauld described two of the streams flowing into Rocky Bayou as being "about 30 Ft. broad and 9 deep, the East branch . . . more rapid." Here too there was cedar as well as cypress.[13] Gauld noted that the seacoast between the entrances of Santa Rosa and St. Andrew bays ran "ESE and SE by E for the space of 52 miles, the Soundings much the same as off Rose Island" and that "the Trees are thick and come pretty close to the shore." On his chart he depicted the red and white hills he characterized as "Hummocks." The reddish ones, he noted, "are remarkable; there are no trees immediately behind them, but green hillocks, which appear pleasant from the Sea." Gauld sought to reassure the navigator approaching this long stretch of coast:

In falling in with the Land, it is easy to distinguish this part of the Coast from Rosa Island, as the hummocks on the latter are all as white as snow, and the trees do not appear in so many Clumps, or Groves, as here: and in making the land to the Westward of Rosa Island, it is easily known by the smooth Beach and the Sounding for there is only 6 or 7 fm. there, at the same distance from the shore, as you have 12 or 13 fm. here, and abreast of Rosa Island.

At a distance of some sixteen miles offshore, about midway between the entrances, "the Land may be plainly seen from the Mast-head of a small vessel," and directly off the two passages into St. Andrew Bay, "the Land may be seen from a Frigate's Mast-head" at a distance of twenty-two miles.

12. Gauld, "A General Description of West Florida, 1769," p. 25.
13. "A Survey of the Coast of West Florida from Pensacola to Cape Blaise . . . 1766."

Coastal Surveying, 1765-1766

Making his way southeastward, Gauld entered St. Andrew Bay, depicting two entrances which eventually would be known as West Pass and East Pass. Separating them was a narrow wooded key which he called "St. Andrew's Island." It too received another name years later: Hurrricane Island. There can be little doubt that *Charlotta* was navigated over the thirteen-foot bar of West Pass to a safe anchorage inside the point of the peninsula. Gauld indicated such an anchorage on his chart, naming the area "St. Andrew's Harbour" and depicting the two masts of a schooner, her pennant flying, her hull obscured by the east end of the "remarkable white sandy hill, or hummock near the entrance of St. Andrew's Bay." This scene he again drew on his chart in a section which he termed "Prospects of several remarkable Parts of the Coast as they appear from the Sea." These ten views or "Prospects" present profiles or elevations of prominent features of this section of the seacoast, including the appearance of the entrances of the four bays as seen from the sea by approaching ships. Before the establishment of lighthouses along the unexplored coasts of the Floridas, when navigation was rudimentary at best, the value of such visual information on a chart was incalculable.

All evidence suggests that Gauld performed his survey of St. Andrew Bay in the cutter while *Charlotta* remained at anchor. He delineated the lower bay and its three arms, presently known as East Bay, West Bay, and North Bay, with a fair degree of accuracy. Obviously considering St. Andrew, as well as Choctawhatchee Bay, to be of little value to the Admiralty because of shallow bars and narrow entrance channels, he gave them only superficial examination, a fact he noted on his chart:

N.B. The interior parts of the Bays of Sta. Rosa and St. Andrew are not laid down with the same exactness as the Sea Coast, but sufficiently near the truth, to give a proper Idea of them. The Method used was, by going round them in a Boat, and, with a common Meridian Compass setting the several Points, and other remarkable Objects, and estimating the Distances.

Apart from "St. Andrew's Island" and "St. Andrew's Harbour" no other geographical features were named. A comparison of Gauld's few notes and modern places and circumstances discloses "A Well" between Alligator and Courtney points and the presence of "Live Oak; Pines and some Cedar; Swampy Ground; Large Pines; Salt

G^{eo.} Gauld — Cartographer

Marshes; and Swampy Salt Marshes" in appropriate places around the shoreline of West Bay. The east side of North Bay afforded "Live Oak and some Cedar," while "Fine Live Oak" was shown in the vicinity of St. Andrew, now part of greater Panama City. As for seafood, only oysters were noted, and these were found at the mouth of Massalena Bayou, near downtown Panama City, and at the entrances of Watson Bayou and the Grand Lagoon. Gauld's close attention to the adjacent Gulf Coast took him ashore, and he found the small "Fresh Water Pond" at the head of the Lagoon near Panama City Beach and the former "Fresh Water" lake through which the present ship channel was dredged in 1934. Davis Point was covered by "A Thick Bluff of Trees, Live Oaks, &c.," and opposite *Charlotta*'s anchorage, on the mainland south of Tyndall Field, Gauld noted the presence of "Large Pines, &c." His soundings were sparse but adequate to show the deepwater trend within the bay and its three arms. Only two shoals were shown: portions of the middle ground between the peninsula and the mainland.[14]

Gauld's written description of St. Andrew Bay added little to the information on his chart. He described the entrance as being "between a Small Island on the Right hand and a narrow Peninsula on the Left." He confirmed the position of the "high white Sand Hill . . . a remarkable object from the Sea," which was some two miles northwest of the entrance, in "Latitude 30°06' N and about 10 Leagues to the NW of Cape Blaise." He described the large shoal guarding the entrance, the channel, and the depth of its bar, but he apparently considered the other entrance between the east end of St. Andrew's Island and the mainland as too shallow to merit description or even mention. The commodious anchorage where *Charlotta* lay afforded a depth of five fathoms and had "the advantage of fresh Water . . . easily got by digging." Gauld's observation that the west arm of St. Andrew Bay extended to "within a few miles of St. Rose's Bay" and that the country between them was "low and Marshy and full of fresh Water Ponds" suggests that he reconnoitered the interior as he surveyed the intervening coast. Bringing his comments on St. Andrew Bay to a close, he noted that it was navigable for any vessel able to cross the bar and that once inside there were depths of three to seven fathoms all over the bay. Finally, he advised that there were no rivers of any consequence and that the soil immediately on the bay,

14. Ibid.; and USC & GS Chart 1263 (9th ed., 1969), *St. Joseph and St. Andrew Bays*.

which was all that he observed, was not to be commended, even though there was a great quantity of large pines, live oaks, and cedar.[15]

The survey from Pensacola to Cape Blaise was more than half finished; only some thirty miles of seacoast, the cape shoals, and St. Joseph Bay remained to be examined and charted. But a problem developed which required immediate attention and which brought Gauld's work to a temporary halt. *Charlotta* was leaking so badly that she had to be careened for repairs. The decision was therefore made to return to Pensacola. The schooner's arrival on August 19 and her condition were described by the master of *Ferret:* "at 10 found the Surveying Schooner working up the Harbour . . . at noon the Schooner came to an anchor found hir Very Leaky in hir Bottom."

The cooper was promptly sent aboard *Charlotta,* and three days later she was brought alongside *Ferret* to transfer her remaining stores in preparation for "heaving down" or careening. The master of the base ship, James Smith, visited the schooner along with the carpenters in order to survey the nature and extent of the necessary repairs. Two days later Smith went ashore to purchase pitch and various small items required for that work. On August 27, *Charlotta* was able to return to her anchorage near *Ferret,* careening and bottom repairs completed. The next three days were spent in caulking her decks. After receiving stores purchased ashore and others transferred from the base ship, on August 31 *Charlotta* moved to an anchorage nearer the watering place at the Red Cliffs; and on September 5 the schooner, with Gauld and his party, departed Pensacola Bay to resume the unfinished survey. It was a short trip. The weather was very "squally," and three days later *Charlotta* was sighted returning to Pensacola. At 2:00 P.M., September 9, she came to anchor, and it was found that she was still "Very Leaky." *Charlotta* was proving to be an expensive and troublesome vessel.[16]

Captain George Murray, commanding *Ferret,* was experiencing problems on his own ship, but of a different nature. *Ferret* had been depleted of manpower, having "lost many Men by Desertion and Death," and it was impossible to secure replacements through the usual channels. Realizing that he might have to take his vessel out on

---

15. Gauld, "A General Description of West Florida, 1769," pp. 25–26. Gauld's latitude was within one mile of the now nonexistent "high white Sand hill."

16. Master's log, *Ferret:* ADM 52/1232.

survey duty in the place of the unreliable *Charlotta,* Murray petitioned Governor Johnstone and his council for assistance. In a letter dated September 1, 1766, Murray informed the governor of his problem and pointed out that as *Ferret* was about to leave Pensacola in order to survey the coast and protect trade, she needed a full crew. This was particularly important, he remarked, "considering the present disposition of the Indians, and the nature of this Service which must frequently expose the Boats to the Danger of being Cut Off." Murray therefore asked for the consent of the governor and council and "the assistance of the Civil Officers in impressing such Seamen as can be found not belonging to any Ship or Vessel sufficient to make up the Complement of His Majesty's Sloop Ferret."

In a hastily called meeting on September 4, at which only three members of the council could be mustered, former navy captain Johnstone put the matter before the board, and a quick decision was rendered. Citing "an Act of Parliament, made in the Nineteenth Year of the Reign of his late Majesty George the Second," wherein it was "declared lawful, in the case of any unforseen necessity" to impress seamen, the governor and council unanimously resolved "that such Consent for pressing Seamen as the Law requires . . . should be given and is hereby declared to be given to The Honourable Captain George Murray, agreeable to his letter aforesaid, upon his Warrant for impressing being indorsed by the Custos Rotulorum of the District."[17]

Being an unemployed seaman in Pensacola was not without its hazards. The record does not indicate how many seamen Captain Murray was able to procure under his authority of impressment, but he exercised his power with some degree of success. A log entry of September 9 indicates that Murray "Sent the pinnace to Town to Impress Some Men, A.M. At 4 She Returned—with one man."

By September 12, *Ferret* had taken fresh water, and with a light breeze she crossed the bar and shook out all sails on course to the southeast, Gauld and the cutter in company. Sounding frequently, her master sighted the "White Clifts" of St. Andrew Bay, referred to by Gauld as a "high white Sand hill," by noon of the second day, bearing E by S ½ S, 2 leagues. Because of light breezes, calm weather, and caution observed during the hours of darkness, it was

17. C.O. 5/632.

the following evening before *Ferret*, guided by the cutter and the longboat sounding ahead, felt her way past St. Joseph Point and halted inside the point with her best bower anchor in five fathoms of water.

Early the morning of the sixteenth, Gauld and Master James Smith entered the boats "to sound on the Bar and Different parts of the Harbor." With the cutter and the longboat both working, only two days were required to complete the soundings and delineation of the bay, and on the morning of the eighteenth both boats returned to *Ferret*.[18]

Gauld reported that the coast between St. Andrew's Island and St. Joseph Bay ran east-southeast about five leagues with a shoal or bank all the way between them. This shoal was of a whitish color and had twelve to eighteen feet on the greater part of it, except near the mouth of St. Joseph Bay, where a bank between the point and the mainland shoaled to seven or eight feet, an area presently known as Bell Shoals.[19] "There is a very good channel with 3 fathoms on the Bar," Gauld wrote:

There requires no other direction to go into St. Joseph's Bay but to keep within about a Cable and a half or two cables Lengths off the Peninsula in 5 or 4½ fathoms; it shoals regularly toward the Point, from which a spit of sand runs out a little way; when in 3 fathoms, hawl round gradually still keeping near two Cables lengths offshore.

"The Bar is narrow, and immediately within it there is from 4 to 6½ and 7 fathoms soft ground," he advised. The end of the peninsula formed "two or three Points, from each of which a small spit runs off for a little distance which may be known by the discolour'd water on them."[20]

As a graphic aid to the seafarer Gauld projected a line on his chart through the axis of deepest water in the entrance channel and over the bar, extending to a prominent tree just to the eastward of a "Clump of Trees" and a "Red Cliff" in the vicinity of modern Beacon Hill. His note on the chart advised: "When you are in the channel coming into St. Joseph's Bay, keep this Tree on with the East end of the Bank [i.e., the shoal which formed the west side of the channel],

18. Master's log *Ferret:* ADM 52/1232.
19. Ibid.; Gauld, "A General Description of West Florida, 1769," p. 26.
20. Gauld, "A General Description of West Florida, 1769," pp. 26–27. A cable's length was 120 fathoms or 720 feet.

till you get into 3 fm. then haul round the Point with an easy Sweep, and you will soon be in 5 & 6 fm." As a further aid Gauld projected another line from the tip of St. Joseph Point through a distinctive grouping of trees on the mainland, just south of present Port St. Joe, near which he noted: "When this Clump of Trees is just on with St. Joseph's Point, you are in the shoalest Water on the Bar," which was seventeen and one-half feet at lowest water.[21]

St. Joseph Bay was described by the surveyor as "an excellent Harbour" with the best anchorage just inside the peninsula. He depicted that area on his chart and characterized the bay as "nearly in the Figure of a Horse Shoe, being 12 miles in length and 7 miles across where broadest; towards the Bottom of it are a few small Islands, and the water is so shoal that a Boat can hardly go near the shore." Fresh water could be obtained by digging close to the old Spanish ruins near St. Joseph Point, and on the north side of the bay were "two or three small fresh water runs" with three or four fathoms close to the shore opposite them.[22]

Gauld offered little encouragement for agriculture, noting that the north side of the bay was very sandy, but near the ruins on St. Joseph's Point there were spots that showed "a kind of verdure and produce plenty of Grapes, some of which are large and of a purple colour and pretty good to the taste." He speculated that they might have been planted by the Spaniards. Naseberries flourished, and a few "small Palmetto Cabbage Trees" were noticed. These grew in abundance on "St. George's Islands beyond Cape Blaise and all along the Coast to the Eastward. . . . The Bud or unform'd Leaves in the Heart of it being boil'd, has somewhat the taste of Cabbage but is more delicious," Gauld concluded, describing a dish which has long been known to Floridians as "swamp cabbage."[23]

If Gauld saw no future in agriculture around St. Joseph Bay, he envisioned a rich harvest of seafood from its waters. Moreover, salt might be made to cure the "Bass, Rock Cod and other kinds of Fish which are here in abundance, and when well cured are little or nothing inferior to those brought from the Northward." On the basis of

21. "A Survey of the Coast of West Florida from Pensacola to Cape Blaise . . . 1766."
22. Gauld, "A General Description of West Florida, 1769," p. 27.
23. Ibid. It is doubtful that the Spaniards took the trouble to plant grapes at a military outpost as insignificant as St. Joseph. Gauld's "naseberries" were the fruit of one of the sapodilla family, probably gum humelia or false buckthorn (*Bumelia lanuginosa*). His "Palmetto Cabbage Trees" were the genus *Sabal palmetto*. See Herman Kurz and Robert Godfrey, *Trees of Northern Florida* (Gainesville, 1962), pp. 23–24, 257–60.

these resources he thought a "very good establishment might be made." On balance, it appears that Gauld considered St. Joseph Bay a body of water of some importance, as did the Spanish and French before him.

Touching on the historical background of the bay, Gauld noted that the Spaniards had erected a post on St. Joseph Point which they abandoned about 1700 but resettled in 1719. The "ruins . . . of the village of St. Joseph" were still in evidence. "In the year 1717 the French erected a Fort Crevecour a Mile to the Northward of a run in St. Joseph's Bay opposite to the Point of the Peninsula," he continued, "but abandon'd it the next year on the representations of the Governor of Pensacola that it belonged to his Catholic Majesty."[24]

With Gauld, the master, and their crews aboard, and his task in and around the bay completed, Captain Murray gave orders to set sail — their next assignment to sound and chart the far-reaching shoals extending from Cape Blaise. By 7:00 A.M., September 18, *Ferret* had cleared the bar and was proceeding on a southwesterly course. Gauld's hydrographic examinations, made with the cooperation of James Smith, required six days to complete, and Gauld recorded soundings as far as sixteen miles off the cape. The task required that Gauld, two petty officers, and four men go ashore to build fires at two designated positions; the smoke from these fires would provide visible points for triangulation purposes and ensure that the distant soundings would be placed correctly on Gauld's chart. During this process *Ferret* remained in the quarter in which Gauld and the master were separately sounding.[25]

Gauld described the peninsula presently known as St. Joseph Spit, the seaward side of which formed the coastline to Cape Blaise, as "a narrow slip of land, in some places not a ¼ mile broad." Gaps in the trees, through which the bay might be viewed from a masthead, together with the "trenching of the Land about NNW and SSE, for near 4 Leagues make it easily known." "A remarkable single Tree" stood apart from the thick trees about the cape and conspicuously marked the point. In an easterly wind, safe anchorage about one or two miles offshore, in six or seven fathoms, was available opposite

---

24. Gauld, "A General Description of West Florida, 1769." Spanish and French rivalry for St. Joseph Bay is discussed briefly in Stanley Faye, "The Contest for Pensacola Bay and other Gulf Ports, 1698–1722," *Florida Historical Quarterly* 24 (1946): 184–85; and in William B. Griffin, "Spanish Pensacola, 1700–1763," *Florida Historical Quarterly* 37 (1959): 254.

25. Master's log, *Ferret:* ADM 52/1232.

the thickest trees. Around the cape, about three or four miles to the eastward, there was a "remarkable gap" in the trees where the bay was separated from the gulf by a "narrow isthmus not above 5 or 600 yards broad."[26]

A comparison of Gauld's manuscript with a modern chart indicates that the bay side of the peninsula has undergone considerable change, while the gulf side has remained very much the same. The surveyor's chart depicts a cove of the bay intruding into the peninsula which probably is today's Eagle Harbor. The "thick trees" near the cape are still present, particularly in the area of Richardson Hammock. The largest of the three islands mentioned by Gauld and shown on his chart is readily identified as Black's Island.[27]

The cape itself was recorded as being in latitude 29°41′ N and ending in a low point about two miles from the trees. Hydrographically distinguishing this landmark was the irregularity of soundings found a great way out at sea. These were marked by several banks of three or four fathoms, six or seven miles offshore, with depths of seven to ten fathoms between them. Finally, Gauld cautioned that there were banks of five to six fathoms farther out, but that there was "no danger in going near enough to make the land plain."[28]

Ferret's underwater exploration with the sounding lead covered an area similar in form to three-quarters of an ellipse, its longest radius extending south-southwesterly from the cape to a point beyond the shoals and the shorter radii formed by the coastlines of the peninsula extending northward and the isthmus eastward. Within this rather large area the depths and bottom characteristics were taken by the base ship and her smaller craft and recorded on the chart by Gauld in considerable detail, especially the locations of the shoals whose shape, extent, and depths he delineated.[29]

Shortly after midday, September 25, the soundings were com-

26. Gauld, "A General Description of West Florida, 1769," pp. 27–28; "A Survey of the Coast of West Florida from Pensacola to Cape Blaise . . . 1766." By "trenching of the land" Gauld meant that the sounding gradients were relatively uniform along the lines of direction and for the distance indicated.

27. Gauld, "A Survey of the Coast of West Florida from Pensacola to Cape Blaise . . . 1766"; and USC & GS Chart 1263, *St. Joseph and St. Andrew Bays*.

28. Gauld, "A General Description of West Florida, 1769," p. 28. Present Cape San Blas lighthouse is in latitude 29°40′ N. See Nathaniel Bowditch, *American Practical Navigator* (Washington, D.C., 1962), p. 1066.

29. Gauld, "A Survey of the Coast of West Florida from Pensacola to Cape Blaise . . . 1766."

Coastal Surveying, 1765–1766

pleted and Gauld returned to *Ferret*. The boats were hoisted, anchor weighed, and sails set for their return to Pensacola. The now familiar landmarks of St. Joseph Point and the "White Clifts of St. Andrew's" were identified and passed at dawn the next day. Topsails, reefed during the night because of squally weather, were shaken out to take full advantage of the wavering winds, but at three o'clock the following morning, black and threatening weather from all points of the compass brought out all hands to double-reef the fore and mizzen-topsails. It was the morning of September 28 before *Ferret* was safely at anchor, the "Garrison of Pensacola N by W, ¾ of a mile." Here they found H. M. S. *Adventure*, the vessel that would serve as Gauld's next base ship.[30]

And what happened to *Charlotta* during the sixteen days of Gauld's absence? Apparently nothing, since she was shifted back alongside the base ship, and the men and provisions aboard her were transferred to *Ferret*. Six days later carpenters were sent aboard *Charlotta* to survey her condition. Their report to Captain Murray was not favorable. She was once again brought alongside *Ferret*, and her ballast and remaining small stores were removed in order "to sell her out."[31] Thus, a vessel which had undergone extensive repairs at considerable cost to the Admiralty and had served Gauld but briefly and unsatisfactorily was put up for sale to the highest bidder.

Once again Gauld was without a suitable ship with which to perform the duties of his assignment to his satisfaction. Rear Admiral William Parry, who had replaced Admiral Burnaby as commander in chief on the Jamaica Station, was so advised. In his letter to the Admiralty accompanying a fair copy of Gauld's most recent surveys, Parry notified Their Lordships of the problem; and although he set about procuring a proper vessel for Gauld, it would be a year before it would be available for his use.[32]

Despite the handicaps incidental to the unreliable *Charlotta*, Gauld had completed his first specifically coastal survey and, according to instructions, sent the cartographic results of his work to his superior

30. Master's log, *Ferret:* ADM 52/1232.

31. Ibid. *Charlotta* was driven ashore in Pensacola harbor and wrecked during the October 1766 hurricane; on July 9, 1767, her sails were stripped for use aboard General Haldimand's schooner. Captain Thomas Fitzherbert to Admiralty, November 1, 1766: ADM 1/1789; captain's log, *Levant:* ADM 51/512.

32. Parry to Admiralty, March 21, 1767: ADM 1/238.

in Jamaica for transmittal to the Admiralty. He had opened to inspection three large bays along this section of the West Florida coast, one of which held reasonable promise should Britain decide to utilize it. Perhaps best of all, the Admiralty had responded favorably to Gauld's initial activity and rewarded him in a tangible fashion. Beginning July 1, 1766, the surveyor's salary was doubled to twenty shillings a day, and the seamen serving with him were to be supplied with a double allowance of grog "to enable them to bear the fatigue in that hot country with more chearfulness and alacrity." The suggestion had originated with George Gauld, and it may be assumed that Jacktar was as pleased by the prospect as was the surveyor by the Admiralty's assurance that certain of his instruments, sent home in care of John Lindsay, would be cleaned, repaired, and returned to him at the first opportunity.[33]

Gauld remained in port less than two weeks before setting out again. Although Captain George Murray had orders to return to Jamaica, both he and the surveyor thought it would be "a misfortune to lose the opportunity of taking the Bar of Mobile" before *Ferret* left the Gulf Coast. As the greater draft of Gauld's new ship, *Adventure*, rendered her unfit for working shallow coastal waters, her captain, Thomas Fitzherbert, was persuaded by Gauld's arguments to allow him the use of *Ferret* in order to survey the coastline westward from Pensacola to the large shoal bay of Mobile. Between the two lay Perdido Bay, so named because of the loss of a Spanish vessel near its mouth. Its entrance was known to be narrow and its bar shallow, so Gauld felt no urgency to give it the rather thorough examination which it later received.[34]

While he attended to the details of *Charlotta*'s final disposition, Captain Murray prepared his own ship for the short voyage to Mobile. Over five thousand pounds of bread were condemned by survey and sent ashore to be replaced by the contractor with a modest amount of fresh biscuit for their current use. Salt beef and pork were likewise conveyed to the ship. Fresh water was replenished, and fi-

33. Admiralty to Gauld, July 12, 1766: ADM 2/726. The surveyor's salary was permanently established at twenty shillings per day on September 17, 1767: ADM 12/55.

34. Fitzherbert to Admiralty, September 25, November 1, 1766: ADM 1/1789; Parry to Admiralty, March 21, 1767: ADM 1/238. Unless otherwise noted, the following events are recorded in master's log, *Ferret:* ADM 52/1232. For the derivation of the name of Perdido Bay and its description, see Gauld, "A General Description of West Florida 1769," p. 17.

nally, on the day before sailing, a small amount of fresh beef was delivered for the ship's company.

Before dawn, October 9, *Ferret* weighed anchor and came to sail. By 7:00 A.M., she had passed Tartar Point and was standing over the bar where she encountered "a Great swell." Bearing away to the westward under full sail and a fair wind, Captain Murray skirted the coast seven to eight miles offshore and anchored off Mobile bar before dark, four miles from the breakers.[35] The forty-one-mile run had taken just ten hours. Their signal for a pilot, repeated one hour later, went unheeded. *Ferret* remained at anchor the next day, and the longboat was launched "to try if a Pilot could be brought on board — but the boat returned at noon, the sea running too high for hir."

The weather, in the meantime, had become hazy, the wind rising at times to strong gales, gradually backing to the north; by early afternoon the following day the swells had increased to "very great from the SE." Taken together, these were sure signs of a tropical storm approaching from the southeast.

The actions of Captain Murray and his master in securing their ship for heavy weather were those of prudent seamen, if not of officers versed in the destructive force of tropical storms. The topgallant yards were lowered and the guns housed. The longboat was hoisted aboard, stowed on the main deck, and fitted with a "wash board" by the carpenters in an attempt to protect it from the heavy seas which *Ferret* might expect.

By the morning of October 12, the northeasterly winds had moderated slightly, but the swells were still very high. The decision was made to heave anchor and attempt to secure a pilot from Dauphin Island. By midday the anchor had been recovered, with considerable difficulty, and was found to have "one arm broke off . . . shank much bent and ye cable much rubb'd." The constant surging from the great swells had taken its toll on their best ground tackle and had almost resulted in their breaking adrift. Once under way, *Ferret* proceeded toward Dauphin Island and by midafternoon had approached to within two miles of Pelican Island, when the pilot boat was seen approaching from the distance. Within the hour Samuel Carr, Mobile pilot, had scrambled aboard the heaving *Ferret*. His boatman, Thomas Harmond, would normally have returned to the pilot station on Dauphin Island, but "a great sea stove and lost the pilot's boat under the

---

35. Gauld noted *Ferret* and her exact position on October 9, 1766, on the initial rough draft of Mobile Bay which he began in February 1768. MODHD E19/10 5k.

lee of the ship in his getting on board." Thus, he too was compelled to come aboard with the pilot.[36]

"Strong gales and a great sea" continued throughout the night, and by morning *Ferret* had "fallen much to the leeward." Faced with the prospect of beating windward back to Mobile Bay and being unable to enter because of high seas, Captain Murray decided to proceed farther to the westward and seek shelter in present Mississippi Sound, under the lee of Ship Island. In these circumstances, Samuel Carr and his man had no choice but to remain as passengers. The reefed foresail was let out and the main topsail set. Driven by strong but fair northeasterly gales, *Ferret* soon passed Petit Bois Island (referred to in the log as "Little Dauphin Island") and Horn Island and by 5:00 P.M., October 13, had anchored in twenty-three feet of water one mile northeast of the western end of Ship Island.[37]

Strong gales accompanied by rain and very hazy weather continued until the early hours of their second day in the haven, but it scarcely mattered; *Ferret* was safely moored, with two anchors, within the lee of the island and secured for heavy weather. As dawn approached, October 15, the wind moderated, and shortly thereafter a boat and crew were sent out to catch fresh fish for the ship's complement. The weather continued to improve, and by midmorning, when the fishermen returned, it was evident that the violent weather had passed. Taking no chances, Captain Murray and his master decided to wait one more day before returning to Mobile.

The calm after the storm, and the light variable breezes, offered a welcome relief, but they did pose certain problems in getting under way and leaving the anchorage. The full extent of navigable water to the north of Ship Island was obviously unknown to Murray, for he sent the pinnace to sound "the different parts of the bay." Finally, the morning of October 17, the light sails were set and *Ferret* got under way, but failing wind brought the order to man the oars and tow the vessel through the narrows at the west end of the island to the safety of the open sea.

Retracing her previous route and sounding hourly throughout the night, *Ferret* arrived off Mobile bar at noon the following day. Samuel Carr was still aboard, but for some reason no attempt was made to

36. *Ferret's* muster book: ADM 36/7376 shows Carr and Harmond aboard October 11 "off Mobile Bar."

37. USC & GS Chart 1267 (14th ed., 1968), *Mississippi Sound and Approaches*, relates this passage and anchorage to modern cartography.

cross the bar and enter the bay. It is possible that Captain Murray intended to remain outside until Gauld had completed his survey and sounding of the entrance and bar. In any event, the ship was anchored seven or eight miles south of Dauphin Island.

By midnight, the first telltale sign of another storm in the gulf was again felt: "A great swell from the S.E." Within four hours it had increased to such an alarming degree that *Ferret* was brought to sail, standing toward the southeast and the relative safety of Pensacola harbor. The wind, meanwhile, had shifted to the east and the master noted in his logbook that "we could not get to windward." Once again they were compelled to run to the leeward to seek shelter. Bearing away toward Pelican Island, they anchored near its west point while the master sounded the swash channel between it and Dauphin Island. He could find only two fathoms, hardly sufficient for safe passage of *Ferret* into Mobile Bay. The longboat was therefore hoisted aboard and the ship stood out to sea.

For the next two days *Ferret* "stood off and on" at the entrance to Mobile Bay while the signs continued to indicate the approach of another tropical storm. The longboat was again launched and a crew sent to sound the main entrance, only to return with the report that they "could not sound the bar, it braking very high from one side to the other." Later the same afternoon, the master himself took the longboat and a crew "to view the Barr; at 5 they returned, found it to brake all over at a surprizing height so that no Vessel or Boat could attempt it."

Other ominous signs were appearing. Not only were the southeasterly swells increasing in size, but the skies took on a black, threatening appearance. The air was heavy with haze, and the increasing wind, backing from east to northeast, indicated that *Ferret* was on or near the track of an approaching storm. In view of their obviously wise decision to run from the storm and seek shelter under Ship Island only days before, one can only speculate on the failure of Captain Murray and his master to pursue a similar course of action now. The inevitable conclusion is that neither was knowledgeable in the ways of tropical storms.

Mobile bar was found to be breaking all over; the order was therefore given to heave anchor and "Runn down the back of the Barr." This took them close to Pelican and Dauphin islands, where they noted "a surprizing swell setting in towards the shore." By now six "Tuns" had been filled with salt water, presumably for ballast to minimize the rolling and racking to which *Ferret* was subjected by the violent and "surprising swells" from the southeast.

As the twenty-second day of October wore on, the weather worsened, and darkness brought no relief. Under much reduced sail, the vessel stood off and on and by noon the next day was in twenty fathoms, an estimated eleven leagues from Mobile bar. Unquestionably, the storm was approaching on a track which would pass to the eastward of *Ferret*. The wind accordingly veered to the northeast, blowing the vessel to sea. Had the storm center passed to the west of the vessel, she would almost certainly have been driven to destruction on the coast.

By midafternoon, all sails having been taken in, *Ferret* was unmanageable under bare masts and was buffeted by "hard storms of wind and rain." Some idea of the violence of the elements and the efforts of the ship's company to save their vessel and themselves is indicated in the terse entries for October 23: "At 4 batten'd down all the hatches and cross lash'd the boats and booms with a hawser—at 6 heavy storms of wind & rain—ship'd much water—both pumps at work." "At 7 P.M. there being 4½ feet water in the hold and the ship laying gunwall under in the sea and blowing a violent storm of wind and rain, cut away the main and mizzen masts with which was lost the whole standing and running rigging with every other material, the mizzen deadeyes excepted." "At 8 got clear of the wreck[age] without the loss of a man—Employed the whole ships company in pumping and bealing [bailing]—violent storms of wind and rain, and a very high sea, at 9, 10, and 11, gained much on the water in the ship." In addition to the main and mizzen rigging and sails, all considered to be the province of the boatswain, dismasting had resulted in the loss of "the whole materials belonging to the carpenter," save the main topgallant yard, main topmast, and two small booms.

The following morning brought no improvement in the weather, but one hopeful sign was noted aboard *Ferret:* the pumps had "sucked," indicating that she was now dry and in no danger of foundering. The crisis had not yet passed, however, and further heroic measures would be needed to save the ship from the force of the hurricane. The log continues:

. . . the storm nothing abated . . . lost the Jibb from the Boom and Main Topmast stay sail—made use of several sails to farther secure the hatches—half past 4 lost the Pinnace off the Boom with all her furniture,

and some of the studding sails — at 5 the storm still continues — Both pumps at work — found much difficulty in keeping the hatches secure — at 8 hard storms of wind and rain — found the foremast to filch much way [to move excessively] *which obliged to cut away the fore topmast* [by which was lost the whole standing and running rigging] . . . *at 9 the storm abated something* — Employed clearing the ship, Lashing the booms and passing rope round the ship — the carpenters employed nailing some board on part of the Quarter Deck that the main mast had torn up.

In all, *Ferret* lost thirteen sails, some in dismastings, some blown away by the storm, some torn to pieces when used to secure the hatches. Cutting away the fore topmast resulted in the loss of the carpenter's materials, but more critical to their safety was the damage to the hull inflicted by the storm and two dismastings: "All the decks topside very leaky . . . four plank on the Quarter deck tore up . . . several of the row ports and all the half ports stove in (by the sea) and broken to pieces . . . all the brick work in the Galley work'd and shaken to pieces."

By noon, October 24, the weather began to moderate, though the sea remained great; the wind, decreasing from strong to fresh gales, soon shifted to a northwesterly quarter, a sure sign that the center of the storm had passed to the eastward, probably not far from *Ferret*. By midnight the skies were clear. No doubt about it, the sloop and her people had survived a tropical storm of hurricane strength, but at grievous cost. Only the lower foremast was left standing — hardly sufficient to maintain way on the vessel, much less navigate her. Murray, his master, and the crew would be forced to draw on their best seafaring knowledge and ingenuity if they were to bring their crippled ship safely into port. As the weather improved, the spare foresail was bent on, set, and trimmed, and a salvaged spar was erected as an improvised or "jury" mizzenmast. The following morning dawned clear, the "sea much fall'n." The ship's company set about rigging a jury mainmast which was fitted with the main topmast yard and its spare sail. With such improvisations *Ferret* was able to wear and tack, albeit in a most awkward manner.

The vessel had been blown some sixty-five miles from her position off Mobile bar to a latitude of 29°02′, according to the master's calculations, and a longitude which was sheer guesswork. In her crippled condition she could only beat her way slowly to windward on a

G^eo. Gauld — Cartographer

course which the captain and master hoped would raise Pensacola bar. Their landfall was actually Mobile Point, and only after further adverse, if not heavy, weather was *Ferret* able to make her way into Pensacola Bay and anchor in the narrows between the west end of Rose Island and the Red Cliffs at 8:30 A.M., October 29. A distress signal was hoisted and several guns were fired. Soon the longboat and pinnace of *Adventure* came alongside to assist in laying out anchors in a futile attempt to warp the ship in to a safer anchorage. Because of a combination of bad weather, contrary winds, and adverse tides, it was almost a week before *Ferret* was anchored a little more than a mile southeast of the "Garrison" of Pensacola. The final stage had been accomplished with difficulty and by means of her own jury-rigged sails, the crew at the ship's oars, and *Adventure*'s small boats towing ahead. Captain Murray celebrated his return by firing twenty-one guns, it "being the 5th of November, the King and Parliament's Delivery from the Gunpowder Plot."

All else aboard *Ferret* was anticlimactic. Massive repairs were begun, and carpenters from *Adventure* were sent aboard to assist. Three men were flogged, one receiving two dozen lashes for "Drunkenness and Mutiny," the others one dozen lashes each for "Drunkenness and Neglect of Duty." In contrast to this harsh treatment was Murray's humane and very practical act of sending seven crewmen suffering from scurvy "to Governor Johnstone's Country House."[38] On November 10, Gauld was detached from *Ferret,* and on the following day he was logged aboard his next base ship, *Adventure,* "per Order."[39]

So ended Gauld's first short and near-fatal voyage to the westward. He had accomplished nothing, but he had learned much. He had experienced the awesome power of a hurricane at sea, and he must have been aware, as never before, of the value of a small, shallow-draft, decked vessel for entering the coastal bays, sounds, and rivers, not only for survey purposes but also when it became necessary to seek a port of refuge.

38. Governor Johnstone's "Country House" was two miles from the fort on the west side of "East Lagoon," now known as Bayou Texar: "A Plan of the Bays of Pensacola and Mobile with the Seacoast and Country Adjacent. By Geo. Gauld M.A. For the Right Honourable The Board of Admiralty 1768"; and USC & GS Chart 1115 (13th ed., 1969), *Cape St. George to Mississippi Passes.*

39. Muster book, *Ferret:* ADM 36/7376; Muster book, *Adventure:* ADM 36/7549. Provided with new masts cut at Pensacola, *Ferret* limped back to Port Royal by February 6, 1767. Parry to Admiralty, March 21, 1767: ADM 1/238.

CHAPTER VI

Coastal Surveying, 1766–1767

His first experience of a hurricane in the Gulf of Mexico might have been expected to make George Gauld welcome the quiet security and relative comfort of a winter ashore, but in little more than a week after returning to Pensacola he was back at sea. The battered *Ferret* would eventually return to Jamaica, as she was ordered to do before her nearly fatal voyage, so Gauld was transferred to H. M. S. *Adventure*'s muster roll on November 10, 1766. The following day Captain Thomas Fitzherbert and George Gauld sailed from Pensacola, intending to investigate the western extremities of the Gulf. Their goal was that great body of water now known as Galveston Bay, but "Instead of going to St. Barnard's Bay . . . as I expected," Gauld reported, "we went no farther that way than to the mouth of the Mississippi, where Captain Fitzherbert wanted to get a Pilot for the Coast if possible, and what Information he could about the Spanish ships that were ashore" as a result of the recent hurricane. *Adventure* stood off the Balize on November 14, hove to, and fired two guns to summon a pilot, "but as it threatened to blow, we stood out to Sea again, when a Pilot Boat was coming out half way from the Balise." Within three hours Fitzherbert was running east before strong gales that split his sails and put an end to any thought of surveying. Steering well clear of the coast, he missed Pensacola and made a landfall between Santa Rosa and St. Andrew bays on the nineteenth. Rather than trying to beat back to Pensacola, Fitzherbert decided to bear away for Havana, "which I was very glad of," Gauld wrote, "having never been there before."

With fair winds, *Adventure* made a good southeasterly run, but as he entered the harbor, November 26, Fitzherbert was greeted by four warning shots from the guns of Moro Castle and forced to await the arrival of a Spanish pilot boat before proceeding to his anchorage. He immediately demanded an explanation of this insult to the British flag, but, like other Royal Navy captains visiting Havana, he got little satisfaction from the Spanish governor. Ostensibly, Fitzherbert put in to Havana because he was short of water; he lay at anchor for a week, replenishing his casks but also taking advantage of the opportunity to observe the reconstruction of El Moro fortress, Spain's Caribbean Gibraltar, which had been besieged and taken by the British in 1762.

During *Adventure*'s stay in Havana, George Gauld "endeavored to make a slight kind of sketch and a view of the new work on top of the hill, opposite the Town, but I had a very poor opportunity of making any remarks [detailed sightings] as no body belonging to His Majesty's ships is permitted to go ashore." Nonetheless, he subsequently sent his sketch to the admiral at Port Royal and noted "a great new Fortification going on . . . behind the Town: there is no judging as yet from the Sea, what figure it is intended to be of, but there is a prodigious number of hands employed about it. They make surprising advances in all their works: we could perceive a considerable difference, even in the few Days the Adventure was there."

Adventure returned to Pensacola on December 11, just one month after her departure. Although her master, Thomas Tripp, picked up a pilot off Santa Rosa Island, he seems to have trusted the most recent navy chart to guide him into the bay, for his log records that "at noon [we] got over Pensacola Barr carrying over 3½ fathoms w[ith] Geo. Gauld's marks on."[1] The surveyor's work was already being put to good use.

During the winter of 1766–67, Gauld turned to the drafting table and prepared "a map of the Coast of Florida from Pensacola to Cape Blaise" which he dispatched to the Admiralty on February 8, 1767, with the observation that it was "the only one I have been able to finish to my own satisfaction. As I have often had occasion to be on this part of the Coast, at all distances from the Land, in men of war and otherwise, I have been enabled to ascertain the soundings even out of sight of Land with tolerable exactness."[2] After nearly two years of hard labor, the surveyor was justifiably proud of his handiwork. He also knew that his efforts were appreciated at home, for he

---

1. Master's log, *Adventure:* ADM 52/1153; Journals of Lieuts. Charles Warburton and Hugh Lawson, *Adventure:* ADM L/A/33 (NMM); Gauld to Admiralty, February 8, 1767: ADM 1/1836; Parry to Admiralty, March 21, 1767: ADM 1/238.

2. Gauld to Admiralty, February 8, 1767: ADM 1/1836. The chart was "A Survey of the Coast of West Florida from Pensacola to Cape Blaise: including the Bays of Pensacola, Santa Rosa, St. Andrew, and St. Joseph, with the Shoals lying off Cape Blaise. By George Gauld, M.A. For the Right Honourable the Board of Admiralty, 1766." MODHD A9464 Press 31c.

had just received the board's letter of July 12, 1766, informing him that his pay had been doubled. Under these circumstances he felt warranted in suggesting to Their Lordships that:

As this present map . . . would be of considerable service to his Majesty's Ships, and others coming upon this Coast, I would beg their Lordships Permission to publish it; and as I have been under particular Obligations to the Earl of Egmont, for using his good offices with their Lordships to encrease my Pay, I should be glad to have their leave to inscribe it to his Lordship, as the first fruits of my Labour, since I have been able to proceed on a less confined Plan than formerly. This is the only Testimony of Gratitude in my power; and I hope his Lordship will allow me that honour.

Looking toward the future, Gauld outlined the ideal plan for properly surveying the Gulf Coast. The whole area should be divided into sections, and a formal survey of each part should be completed before passing on to the next, "for a great deal of time is lost in skipping from place to place." He would begin by charting the coast between Pensacola and the Mississippi, then the sectors from Cape Blaise to Apalachee, from there to Espíritu Santo, thence to the Martyrs or Keys. "The Tortugas and Martyrs would make a large and very Material Division" by themselves. After thoroughly surveying these British shores, "I should be glad to take a Cursory sketch of the Coast west of the Mississippi."

The execution of his scheme, or any part of it, clearly required "a good stout vessel." *Charlotta* was "so bad that it will be impossible ever to do any thing with her at sea." He had urged Captain Fitzherbert to get her in shape for inshore work west of Mobile Bay by March or April, but his problem would best be solved by the acquisition of a new sixty-ton vessel that drew no more than seven feet of water and was capable of stowing two good boats, for "a great deal of Business must be done by their Means." A schooner would be better than a sloop, "as the sails are more easily managed." She must be well supplied with cables, anchors, and pumps in order to operate safely on the open coasts of West Florida. The surveyor and the two boats would often be absent from their ship, so "there should be a sufficient number of hands to manage her exclusive of the Boats Crews. I think about 24 or 25 hands including a Petty officer or two, myself and servant, would be sufficient." Gauld also suggested that the survey ship should be victualed by the agent victualler at Pensacola rather than from naval vessels in the harbor, because "the Pursers do not think themselves obliged to furnish Candles and other necessaries." Furthermore, ships' pursers regularly withheld a percentage of all supplies — "the Purser's Eight" — which meant that on a long voyage crews ran short of provisions. If such an arrangement should be approved, the surveyor himself offered to act as purser for his ship.

Accuracy in surveying depended on precision instruments, and Gauld expressed delight at hearing that his original set, carried home by Sir John Lindsay, would be returned to him. He particularly needed the telescope, in order to establish longitude, and "the Case with the Proportionable and Triangular Compasses." The astronomical quadrant, he suggested, should first be sent to its maker, Mr. Dolland, to be repaired; it was a very good instrument but too delicate. The surveyor advised the Admiralty that he had ordered a good watch with a second hand from Mr. Cumming, watchmaker in Bond Street, and had a standing order with his bookseller for "Monsr. de la Caille's Tables, and de la Lande's Ephemeris always as soon as it is published." The theodolites were "still in pretty good condition, and I hope they will last out this Service."

Turning to another consideration, Gauld asked that he might have John Forbes, midshipman aboard *Ferret,* as one of his surveying party.

He was with me last Summer, in the little Schooner [Charlotta], and was of great service, both in taking angles with one of the Theodolites, for laying down Shoals &c, and otherwise as occasion required. He is a good Seaman, and a brisk, active, careful person, and has been a good deal used to a small vessel, since he came out here, Midshipman in the Tartar. He is particularly acquainted with the Coast and Islands, between Mobile and New Orleans, having had the command of a small Schooner, while the Tartar was here, cruising in those parts, and in the Lake Ponchartrain, for smugglers, and made several pertinent Remarks on the Coast during that time.

Forbes seemed willing to join the surveyor if such duty might be counted toward fulfilling the customary seven years a midshipman must serve before being examined for a lieutenant's certificate; he had one more year to serve and did not wish to lose the time. Gauld also urged that the young man be offered an allowance of five shillings a day "for his trouble, which is really necessary in this Country, where every thing is so immensely dear."³ The surveyor also sought Admiral Parry's support in this matter, but in spite of Gauld's assurance that an assistant would "forward him greatly in Surveying the Coast," Parry declined to take responsibility for an additional salary and merely referred Gauld's request to the Admiralty. There, predictably, it disappeared.⁴

Some assistance Gauld did acquire, however. On December 31, 1766, the name of Peter Quash, the surveyor's "servant," was entered in the muster books of H.M.S. *Adventure*. Like Gauld, Quash was carried as a "supernumerary for victuals only," and for the next ten years his name would follow that of George Gauld in the muster books of every ship to which the surveyor was assigned. Quash was a native of Jamaica; that and his peculiar name strongly suggest that he was a Negro or mulatto, but whether slave, free, or indentured is unknown. Whatever his race or condition, Peter Quash must have been a man of remarkable character and quality to have served — and survived — so long and so faithfully. Quash's service ended September 17, 1776, when he was discharged "by request."⁵

Even a single servant who could perform household chores would have added considerably to George Gauld's standing in Pensacola and to the comfort of his daily life. By the spring of 1767, Gauld had probably built a house, and with an annual salary of £365 he was a man of substance in the young British colony. He enjoyed a good reputation among the merchants, officials, and military officers who made up Pensacola society, as later developments would indicate. The only specific evidence of his role in community life at this time is his signature as a witness to indentures dated March 2 and 3, 1767, whereby, for the sum of £2,000, Joseph Garrow purchased a house and property from James Bruce.⁶ Association with Bruce, a prominent landowner and member of the council, indicates that the naval surveyor moved in the best circles.

The coming of spring brought an end to West Florida's changeable winter weather, much of which had been distinctly unpleasant, for Pensacola had experienced sleet and a rare snowfall in February.

3. Ibid.

4. Parry to Admiralty, March 21, 1767: ADM 1/238.

5. Muster tables, *Adventure:* ADM 36/7549; *Sir Edward Hawke:* ADM 36/7256; *Florida:* ADM 36/7978.

6. C.O. 5/601: 279–83.

Spring also witnessed the arrival of General Frederick Haldimand, the newly designated brigadier of the southern district of North America. Borne on the sloop *Cygnet*, he received a thirteen-gun salute from *Adventure* as his vessel anchored off the fort March 24, 1767.[7]

The arrival of this competent Swiss mercenary had an immediate impact upon life in Pensacola.[8] Horrified by the conditions he found in and around the crowded fort, Haldimand set to work building new barracks, strengthening the defenses, improving sanitation, and clearing the open space between the palisade and the surrounding town — the ground upon which George Gauld's house fronted. No less important was the brigadier's establishment of harmony between the military and the civil authorities of the colony, for Pensacola bred disputes as rapidly as mosquitos. The departure of Governor Johnstone in January had eased the situation, but Haldimand faced the task of restoring order, and that included resolving the charges and countercharges of two of his subordinates, Major Robert Farmar and Lieutenant Philip Pittman.

Philip Pittman was a young surveyor-engineer, detached from the 15th Regiment, who, like Major Farmar of the 34th Regiment, had come to West Florida in 1763. Pittman had busied himself professionally at Pensacola and on the Iberville and had ascended the Mississippi with Major Farmar's expedition of 1765 to the Illinois country. He and Farmar had quarreled violently, and Pittman had associated himself with Governor Johnstone in bringing charges of misconduct against Farmar. Both officers returned to Pensacola in 1767 in anticipation of Farmar's court-martial, and Haldimand found it necessary to adopt every means of keeping them apart. The devastation wrought by a winter storm at Fort St. Marks, Apalachee, offered an opportunity to put the lieutenant's talents to work and to get him out of Pensacola. On June 5, Haldimand ordered Pittman and a carpenter to examine the damage, survey the environs of old San Marcos de Apalachee, and investigate land communication between that isolated post and both Pensacola and St. Augustine.[9] As their

---

7. Master's log, *Adventure:* ADM 52/1153.

8. See Robert R. Rea, "Brigadier Frederick Haldimand — The Florida Years," *Florida Historical Quarterly* 54 (1976): 512–31.

9. See Robert R. Rea, Introduction to Philip Pittman, *The Present State of the European Settlements on the Mississippi,* facsimile ed. (Gainesville, Fla., 1973), pp. xi–xxxv. On Fort St. Marks see Mark F. Boyd, "From a Remote Frontier," *Florida Historical Quarterly* 19–20 (1940–42).

trip could best be made by sea, the Navy was called upon to provide support for the little expedition.

In addition to bringing Haldimand to Pensacola, H.M.S. *Cygnet* had also brought a new vessel for George Gauld's use. Being "uneasy that the Survey should be Retarded, the Government put to so great an Expense, & Mr. Gauld obliged to stop his proceedings for want of a Vessel both safe and convenient to carry his operations into execution," Admiral William Parry at Jamaica had purchased "a large Deal cutter, thirty feet in length, and eight in breadth — almost new . . . well fitted in every respect, with Oars, Masts, sails &c." "She sails & rows extremely well," Parry remarked, and he hoped she would "keep Mr. Gauld employed 'till I can send him down a proper Vessel." This sloop had been entrusted to Captain Durrell of *Cygnet* for delivery to Fitzherbert of *Adventure* and was at Gauld's disposal in April and May.[10]

On June 1, Captain Basil Keith, commanding H.M.S. *Levant*, took up his station in Pensacola harbor, relieving Captain Fitzherbert and *Adventure*. George Gauld was duly transferred to the newly arrived vessel, "per order of Rear Admiral Parry," and along with the surveyor went the cutter which would serve him for the next two years. Like her predecessors, Gauld's new base ship was unsuitable for close coastal surveying and would remain at anchor in Pensacola Bay for the six and a half months he was borne on her muster table, but her spare boat would exactly fit the needs of an assignment such as Pittman's — if the navy surveyor would agree to its use.[11] George Gauld doubtless leaped at the chance to extend his eastward investigation some eighty miles beyond Cape San Blas and into Apalachee Bay, and for once he would have the companionship of a man who shared his professional competence and interests.

Arrangements were accordingly made; by the second week of June all was in readiness, and the two surveyors received their instructions. Haldimand advised General Thomas Gage that "Lt. Pittman departed the 12th for Appalachy with Mr. Gall [*sic*] (a surveyor employed by the Admiralty to draw the plan of the coasts.) I trust that they will return shortly," continued the brigadier, and he promised to forward to army headquarters in New York "an exact report of that post." Captain Keith's log reflected the navy's view of the ex-

10. Parry to Admiralty, March 21, 1767: ADM 1/238.
11. Master's log, *Adventure:* ADM 52/1153; Muster book, *Levant:* ADM 36/7626; captain's log, *Levant:* ADM 51/512.

pedition: "A.M. the 12 Oar'd Cutter man'd; sailed with the surveyor to survey the coast."[12]

No record of the voyage from Pensacola to St. Marks and return has survived, but it is likely that the journey was pleasant enough — Pittman concluded that it was happier to sail along the coast than to traverse the pine barrens and ford the rivers between Pensacola and Apalachee. Gauld would have pumped the army lieutenant for information regarding the western waters of British Florida, the lakes and rivers on which the military surveyor had spent many months and which he had sketched with commendable accuracy. Pittman would have waxed eloquent over his Mississippi River experiences and the superb draft of the river's course which he had made. Without any doubt, Philip Pittman did most of the talking — and his canny Scots colleague proved to be a good listener.

Gauld and Pittman proceeded directly to their destination, for the navy surveyor had traced that coast only the previous year. They conducted their examination of St. Marks in an expeditious manner and returned to Pensacola after an absence of just one month. Gauld's subsequent surveys of the coast of Apalachee Bay indicate that in 1767 he lacked the time to satisfy his requirements, and expectation that he would one day return would have relieved him of any desire to tarry. Nor could the cutter have carried sufficient supplies for an extended voyage. The four-hundred-mile round trip would have taken slightly more than eight days, at an average speed of four knots, if they followed the usual custom of anchoring along the coast at night. Their sojourn in "Apalachy" would then have been about twenty-two or twenty-three days.

It may be assumed that upon arrival at Fort St. Marks, Gauld and Pittman divided their forces and each began the duties peculiar to his own service. Pittman remained at the fort some time in order to examine its condition and confer with Ensign James Wright, the only officer in the tiny outpost. He then set out to the northward on a triangular route which took him to "Tallahassa or Tonaby's Town" and to "Mikisuki or Newtown, Indian Village," thence back to the fort. Gauld, on the other hand, would have returned to the mouth of the river, which he sounded and sketched. Proceeding upstream he examined and traced St. Marks River and its northwest branch, the Wakulla. From their month-long investigations Gauld and Pittman

12. Haldimand to Gage, June 16, 1767 (in French): Haldimand Transcripts, Public Archives of Canada, Ottawa. Captain's log, *Levant*, June 13, 1767: ADM 51/512.

compiled a cartographic record of their findings. Unfortunately, their joint authorship tends to obscure the precise nature and degree of contribution of each man.[13]

From Pittman's hand alone came four signed but undated reports to General Haldimand: "A Description of the Face of the Country," "A Description of the Fort at Apalachy and the Lookout Tower," "Communication from Pensacola to Apalachy," and "Communication from Apalachy to St. Augustine." Although these reports were written by Pittman, a part of the first, wherein he indicated the latitude and briefly described the channel entrance and the rivers, may have been provided by Gauld. The description of the fort, fairly technical in nature and exact in measurements, is clearly the work of a military engineer. In closing his report on "the Face of the Country," Pittman declared that "these descriptions are accommpany'd by a general sketch of the country as far as I saw it," thereby confirming that they were written on the basis of his own observations. His reports on the trails from "Pensacola to Apalachy" and thence to St. Augustine were obtained "by interrogating some Indians and Traders who have travell'd these roads" and "from the Journal of a Gentleman who had travell'd the road."[14]

The Gauld-Pittman "Sketch of the Entrance from the Sea to Apalachy and part of the Environs" depicts soundings, the least of which was seven feet, along a broken line projected north-northwest across the end of modern Sprague Point. The seamark was "The Lookout Castle," near which were grouped four buildings described as "Frame Quarters." The entrance, narrowing sharply near Three-mile Point, and the St. Marks River are sketched to the confluence with the Wakulla and some eleven statute miles beyond. Here a note indicates that the river was "400 yards across . . . overgrown with long grass, bull rushes and angelica, and is no farther navigable." The course of the Wakulla is outlined for some eight miles — farther than its actual length — beyond its junction with the St. Marks, where it was "overgrown with bull rushes, thus far navigable for canoes." The

13. These inferences are implicit in "A Sketch of the Entrance from the Sea to Apalachy, and part of the Environs taken by George Gauld Esq. Surveyor of the coast, and Lieutenant Philip Pittman, Asst. Engineer," Gage Papers, William L. Clements Library, Ann Arbor, Michigan.

14. In Haldimand to Gage, August 5, 1767: Gage Papers (American Series), William L. Clements Library, Ann Arbor, hereafter cited as Pittman, "Reports and Descriptions." Reproduced in Mark F. Boyd, ed., "Apalachee during the British Occupation," *Florida Historical Quarterly* 12 (1934): 114–22.

branching of the St. Marks and the Wakulla is labeled on the sketch, appropriately enough, as "N.E. River" and "N.W. River," respectively.[15]

Of particular interest to George Gauld and any seafarer entering the St. Marks estuary was the seamark, a "small castle," actually a twelve-foot-square tower, forty-five feet high, located on the opposite side of the river from the fort, bearing SSW ½ W, 1,018 yards from "the Angle of the Bastion of the Fort." Describing this tower, Pittman noted that the walls were "2 feet 10 inches thick and pierced with loopholes on each side," thus inferring that, like the fort, it was constructed of stone. Pittman thought that the "castle was design'd as a defence to the workmen employed in the stone quarries, which are within a few yards of this building." He found that it was in great disrepair, "wanting stairs to ascend the Tower, and some joists to the officers room which have been cut away and all the planks of the flooring carried off."

Pittman's description of the fort and lookout tower provides their geographic location and succinctly outlines the entrance from the sea and the course of the river:

The Fort of Apalachy is in the Lat. of 30 d 10 m North and about 237 miles E. of Pensacola and 213 miles W. of Saint Augustine, is situated on a small point of land form'd by confluence of two small rivers which run from the N.E. and N.W. They mix their waters at this point and become one channel of about 900 feet broad running S. 9 miles, and then empties itself into the Bay of Mexico. The Navigation of this Channel for upwards of 4 miles is so intricate owing to a vast number of oyster banks, which form a labarynth that it would require at least a months examination before one could attempt to give any just account of it, however, vessels that draw 7 feet water may go safely to the warf at the Fort.

The fort, buffeted by winds and floods, was in a sorry state of repair. Originally designed for at least two hundred men, it had not been completed by the Spaniards before the loss of Florida to the British in 1763.[16]

Along the "roads" to the interior depicted on the Gauld-Pittman "Sketch" are certain features intended, according to Pittman, to "clear up whatever may be wanting" in his descriptions. Near the

15. Gauld and Pittman, "Sketch of the Entrance from the Sea to Apalachy"; and USC & GS Chart 484 (3d ed., 1965), *St. Marks River and Approaches.*
16. Pittman, "Reports and Descriptions."

fork of the road leading northward from the fort were two wells, one "in the solid rock 6½ fathoms deep and about 30 feet in diameter and 20 from the surface of the ground." A third well is shown hard by the road halfway to "Tallahassa or Tonaby's Town," and a swath of "hurricane ground" appears some three miles farther on. Three ponds are depicted along the northern portion of the road. About six miles south of the Indian village the road branched to the northwest, ending at "The Old Spanish Fort," no doubt the "ruins of San Luis Fort and Town," known to have been some two miles west of Tallahassee and six miles east of the Ochlockonee River. Gauld and Pittman's "Tallahassa" was situated on a southwesterly flowing stream — shown later as the Ochlockonee — some four miles north of modern Tallahassee. Their "Sketch" appears to be the first to record the name of the state capital if not its actual location. In referring to the "N.W. River," depicted as the Wakulla, Pittman erroneously related that "its source is 25 miles N.W. of the Fort [St. Marks], and near where a Spanish Fort [San Luis] formerly stood." He mistook the seven-mile river of spring-fed origin for the Ochlockonee, which takes its rise in southern Georgia.[17]

Along the road from "Tallahassa" to "Mikisuki or Newtown, Indian Village" were noted large and beautiful savannahs with "old fields" in between. Features along the road from St. Marks to "Mikisuki" included pine barrens, "Old Fields," cypress swamps, and a savannah where the "Path to St. Augustine" crossed. This was no doubt part of the route which came to be known as the Old Spanish Trail. Reference to it as a "Path" was probably apt, for all of these "roads," were nothing more than very slightly improved Indian trails. About nineteen miles from the fort, Gauld and Pittman indicated that "the land begins to be pretty good." Completing the toponymy of

17. Ibid.; Gauld and Pittman, "Sketch of the Entrance from the Sea to Apalachy"; *U.S. Coast Pilot 5*, p. 95; Mark F. Boyd, "Mission Sites in Florida," *Florida Historical Quarterly* 17 (1939): 264–69. The exact location of Fort San Luis was known to early Florida settlers, surveyors, and mapmakers who viewed the ruins before vandals, time, and the elements erased the last visible evidence. The site has been verified by documentation and archaeological examinations. See Mark F. Boyd, Hale G. Smith, and John W. Griffin, *Here They Once Stood* (Gainesville, Fla., 1951), pp. 1–4, 139–44, Plates 1, 4; Bertram H. Groene, *Ante-Bellum Tallahassee* (Tallahassee, Fla., 1971), p. 10. The Gauld-Pittman "Sketch" was projected on Gauld's "An accurate chart of the Coast of West Florida and the Coast of Louisiana," published by Faden in 1803, with additional and expanded notations which confirm "Tallahassa" as being on "A Branch of the River Okalakana" (Ochlockonee).

the "Sketch" is a short segment of the "Road to St. Augustine" projected eastward from Fort St. Marks.[18]

With the completion of their separate surveys, reports, and joint sketch, the surveyors' work in "Apalachy" came to an end. In anticipation of their departure, Ensign Wright prepared on July 9 dispatches for transmission to Brigadier Haldimand. Wright advised his superior that one of the St. Marks garrison force was so ill with a "Cairus [carious] bone" that the local doctor despaired of curing him with the medicines at hand. Fearing that the inevitable delays in receiving additional supplies from St. Augustine would prove fatal to him, Wright was sending the soldier back to Pensacola with "Mr. Gall." The party must have left shortly thereafter, for on July 14, the watch aboard *Levant* reported that "at noon our 12 oar'd cutter returned with the surveyor."[19]

Apart from his sojourn at St. Marks and Apalachee with Philip Pittman, the chronology of George Gauld's activity while attached to *Levant* is obscure. Analysis of his successive charts of the coast and land surrounding Pensacola does suggest that in 1767 he spent a considerable time investigating nearby Perdido Bay. Gauld's examination of this relatively unimportant body of water and its tributary rivers must have been made some time after May 6, 1766, when his draft of a "Plan of the Bay of Pensacola, and a sketch of the Environs" was transmitted to the Board of Trade with Governor Johnstone's glowing letter of praise. Since Perdido Bay was fully depicted on Gauld's "A Plan of the Bay of Pensacola and Mobile with the Seacoast and Country adjacent . . . 1768," it follows that it was surveyed within this time span and probably in the latter part of the year 1767. Because of its proximity to Pensacola and because its shallow bar imposed severe limitations upon the size of vessels which might enter it, Gauld probably utilized the cutter that was now available to him to visit Perdido Bay when weather and circumstances permitted. In any case, by 1768 the surveyor had explored and charted this first bay west of Pensacola and had extended his investigation well into the "Country adjacent."

The Perdido River and its estuary, Perdido Bay, opened into the

18. Gauld and Pittman, "Sketch of the Entrance from the Sea to Apalachy."
19. Wright to Haldimand, July 9, 1767: Haldimand Transcripts; Mark F. Boyd, "From a Remote Frontier," *Florida Historical Quarterly* 20 (1942): 395; Captain's log, *Levant:* ADM 51/512.

sea about four leagues west of Pensacola and ten leagues east of Mobile Point. The narrow Perdido entrance, with a depth of only six feet on its bar, widened and extended "upward of a league" to the northeast — to within a mile of Pensacola's Great Lagoon, a common portage for canoes — and thence westward for three or four miles, where it formed a large bay. This in turn extended to the northeast for more than five leagues, with a width of from four to six miles except at one point where it narrowed to a mile, thereby providing Gauld with the appellations of "Lower Bay" and "Upper Bay." The west side faded into a large lagoon. Gauld noted on his chart that "This lagoon is left unfinished for the present: but it is said to extend near as far as Bon Secour," a river running into Mobile Bay. The soundings of Perdido Bay indicated on the chart show rather uniform depths of nine to twelve feet throughout, and Gauld depicted the "Red Cliff" presently known as Red Bluff, on the west side of the bay at the entrance to Soldier Creek.[20]

"There are several considerable runs that fall into the upper part of this [bay], in one of which a saw mill is lately erected," Gauld wrote. Drawing upon his Banffshire origins for inspiration, he named these streams on his chart the Dullen, Fiddich, and Isla rivers. The first two joined about two miles from the bay to form a stream now known as Eleven Mile Creek. "Tate's sawmill" was located on the Dullen about one mile from its confluence with the Fiddich, and a road was traced from the mill to Pensacola. This sawmill undoubtedly belonged to David Tait, sometime deputy surveyor, Indian commissioner, and author of a most interesting journal of his travels among the Upper and Lower Creek Indians in 1772. Tait had received a grant of land on January 10, 1767, and at a meeting of the governor and council on November 5, 1768, he was granted a thousand acres on the Perdido River on condition that it be settled and developed within two years. Tait seems to have anticipated the grant and to have imported six workmen to erect his sawmill.[21]

The Perdido River was described as entering the northwest corner

20. Gauld, "A General Description of West Florida, 1769," p. 17; "A Plan of the Bays of Pensacola and Mobile with the Seacoast and Country adjacent, by Geo. Gauld, M.A. For the Right Honble. The Board of Admiralty, 1768," MODHD, D964, Rt; USC & GS Chart 1265, *Pensacola Bay and Approaches.*

21. Ibid.; Cecil Johnson, *British West Florida, 1763–1783* (New Haven, 1943), pp. 130–31; Clinton N. Howard, *British Development* (Berkeley, 1947), pp. 95, 105n.32; "Journal of David Taitt's Travels from Pensacola, West Florida, to and through the country of the Upper and Lower Creeks, 1772," in *Travels in the American Colonies*, ed. Newton D. Mereness (New York,

Faden in 1803, and by the Admiralty in 1823. As far as a single example can, it epitomizes George Gauld's work as naval surveyor on the Gulf Coast. (Courtesy of the Library of Congress.)

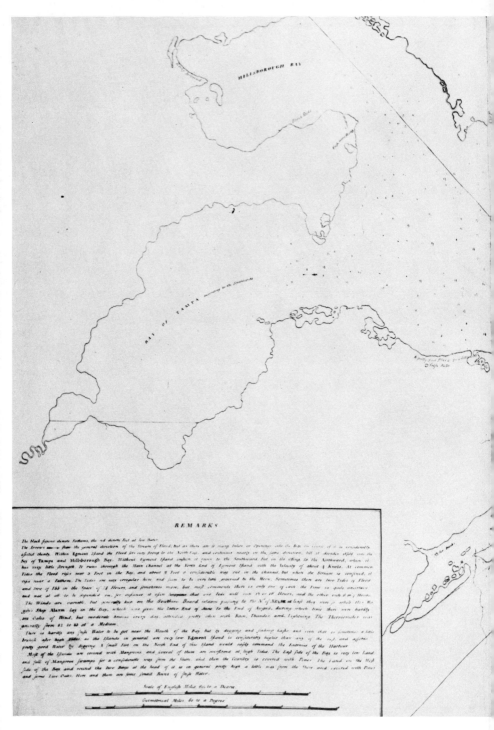

V. "A Survey of the Bay of Espiritu Santo in East Florida . . . by Order of Sir William Burnaby." Gauld's objective in 1765 was the charting of Espiritu Santo or Tampa Bay. The small islands at the entrance are Egmont *(left)* and Burnaby *(right)*. Soundings

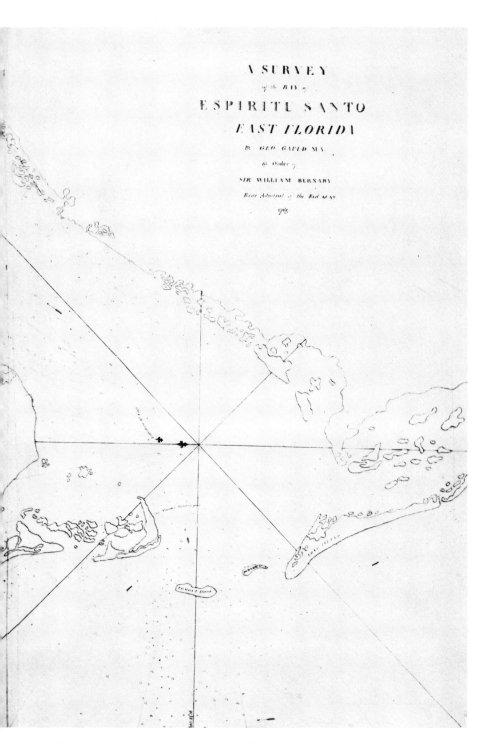

mark the channels and lower reaches of Hillsborough and Tampa bays. (Courtesy of the Ministry of Defence Hydrographic Department.)

VI, VII & VIII. The Coast of the Gulf of Mexico. This is a portion of the great *Chart of the Coast of West Florida, and the Coast of Louisiana,* which was first published by

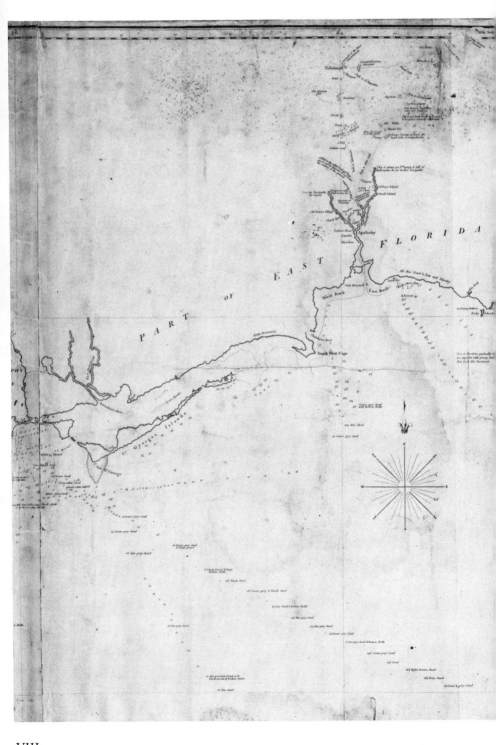

VIII.

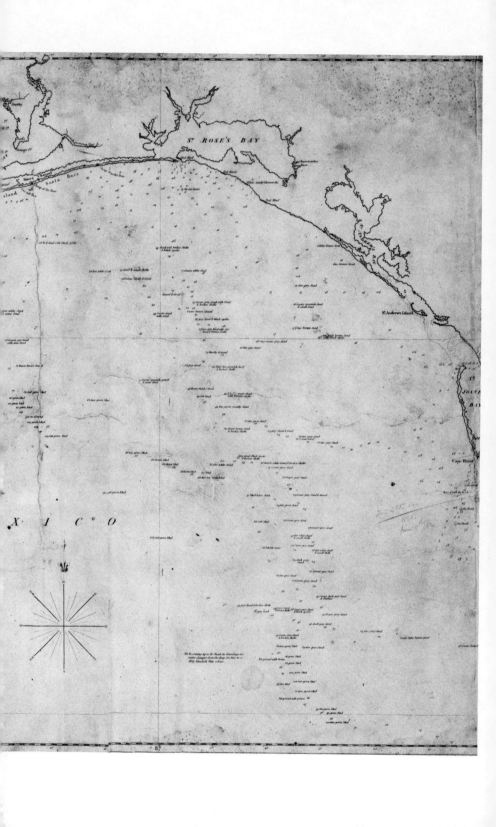

VII.

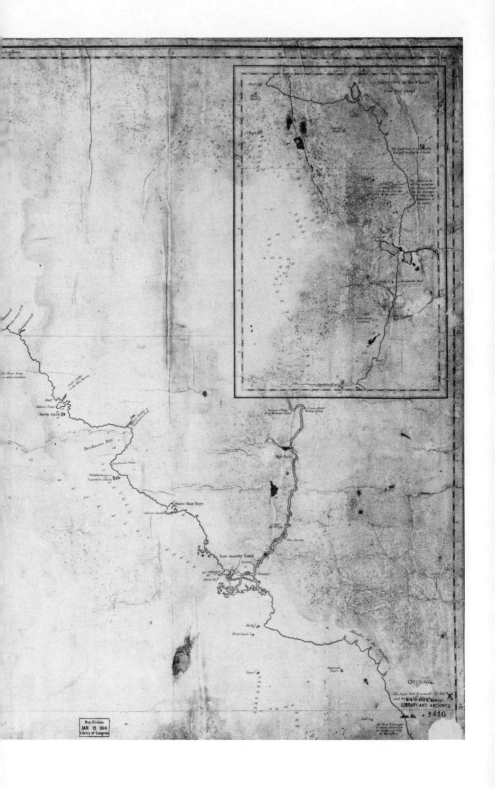

of the upper bay amid "a number of swamps and broken Islands"; it extended in the general direction of north-northwest and had a breadth of 80 to 160 feet — "a pretty rapid stream," in the surveyor's judgment. Its source was supposed to be eighty or ninety miles from the bay, Gauld added. He ascended its winding course about twenty-two miles, as the crow flies, which took him to a point near an Indian camp some two miles above the "Logs across the Perdido," the improvised crossing of the "Old Road" to Mobile. This road, probably little more than an Indian trail, ran north-northwest from Pensacola fourteen miles, crossing the headwaters of several small branches of the Escambia River, then turned westward and crossed the Perdido River at the log ford some twenty miles above its mouth. About three miles before the road reached the east side of Mobile Bay it divided, the northern fork terminating at the plantation of Augustine Rochon, approximately where today U.S. Highway 90 and Interstate 10 reach the bay. The southern fork ended at a "village" on the shore some five miles down the bay. The overland journey between Pensacola and Mobile was inconvenient and circuitous at best.[22]

"In some places, the river is entirely choaked up with logs: the banks are cover'd with cedar, cypress, &c., and the soil in many places is tolerably good for culture but in general it is all good pasture ground," Gauld wrote. He added a historical touch when he noted that the Perdido River was formerly the boundary between Spanish Florida and French Louisiana. Gauld sounded the river as far as he went and recorded on his chart certain nearby topographic features: "swampy land," "white bank," "red bank," "high land," "pretty good soil," and about one mile south of the road, on the southwesterly bank of the river, a "Stony Mt."

The "New Road" to Mobile is indicated on Gauld's chart of 1768 as having been completed from Pensacola to the narrow part of Perdido Bay, between present Cummings and Chagrin points, and a mile or so beyond. Following a fairly direct route west from Pensacola, the road was served by a ferry at the narrow part of the bay. First proposed in a council meeting in 1764, it was not until the General Assembly of 1767 that a bill was offered in the lower house "for applying certain sums of money for establishing a ferry at the River

---

1961), pp. 493ff.; USC & GS Chart 1265, *Pensacola Bay and Approaches;* USC & GS Chart 1266 (23d ed., 1969), *Mobile Bay.*

22. Gauld, "A General Description of West Florida, 1769," p. 17; Gauld, "A Plan of the Bays . . . 1768."

Coastal Surveying, 1766-1767

Perdido and towards a road from Pensacola to the Bay of Mobile." The bill was passed by the assembly and approved by Lieutenant Governor Montfort Browne, and the route was surveyed by Elias Durnford. In "A General Description of West Florida, 1769," Gauld noted, "Since that was wrote the ferry has been established and the new road made, which is not only better than the old road, but cuts off a space of near 20 miles. The journey is now performed with ease in one day which formerly took up almost two days."[23]

The new road was a sign of the slow but persistent development of British West Florida. By 1769 George Gauld's involvement would be far more than that of mere naval surveyor; having charted the colony's coasts, he would help to chart its progress.

23. Ibid. Howard, *British Development*, pp. 36-37, 137n.37, speculates that the designation "New" might have referred to an improvement of the old road; in fact it was a new and different route. See also Robert R. Rea and Milo B. Howard, Jr., *The Minutes, Journals, and Acts of the General Assembly of British West Florida* (University, Ala., 1979), pp. 82ff.

CHAPTER VII

Surveys to the West, 1768

On November 7, 1767, the armed schooner *Sir Edward Hawke,* Lieutenant Charles Warburton commanding, arrived at Pensacola. Admiral Parry, mindful of George Gauld's earlier recommendations and recognizing his need for a relatively small, shallow-draft base vessel, had done his best to fulfill the surveyor's desires. Before *Hawke* could be put into service, however, certain minor refinements and modifications would have to be made, including a complete overhaul of her canvas. With Captain Keith's cooperation, the carpenters and sailmakers of *Levant* spent the better part of the next two weeks attending to these important details.[1]

The major changes on the surveyor's new ship had been completed on December 21, when Gauld was detached from *Levant* and his name and that of his servant, Peter Quash, entered on the muster table of *Sir Edward Hawke.*[2] For reasons that are not clear, the vessel remained at anchor at Pensacola until the latter part of February; perhaps her inactivity was due to the requirements of Gauld's work ashore, but more likely Gauld was well advised to remain in harbor during the winter's intermittent bad weather. A particularly violent rain squall, with thunder and lightning, had struck the harbor on January 9, 1768, capsizing several ships but leaving *Hawke* undamaged. Lieutenant Warburton had dispatched all of his boats to assist the distressed vessels and their crews. Among those "overset" was the governor's schooner, which was righted by Warburton's men. Early in February, some two weeks before she was due to sail, *Hawke* was heeled, and her crew "pay'd the Bottom with Pitch and Turpentine" to protect her from the ubiquitous worm.

By February 20, all was in readiness. With Gauld and his survey party aboard, *Hawke* crept slowly down the bay from her anchorage, but the breeze failed and an impenetrable fog settled heavily over

---

1. Captain's log, *Levant:* ADM 51/512; Parry to Admiralty, October 30, 1768, ADM 1/238; captain's log, *Sir Edward Hawke:* ADM 51/4342. Unless otherwise noted, the record of the activities of Gauld and the survey vessel *Hawke* is derived from this last source.
2. Muster table, *Sir Edward Hawke:* ADM 36/7256.

Pensacola. Lieutenant Warburton was forced to wait until the twenty-fourth, when rising winds and morning sun cleared the air, enabling him to come to sail at eight o'clock. Crossing Pensacola bar an hour later, he stood along the coast toward Mobile, taking regular soundings of four to five fathoms one mile offshore. With the wind over her stern and all sails set, *Hawke* made the forty-six-mile run to an anchorage off Dauphin Island at a creditable speed of almost five and one-half knots. By the twenty-ninth, Warburton had found a snug berth between Pelican Island and the east end of Dauphin Island. Inasmuch as his ship remained at anchor for three weeks while he worked, the protected cove in which she lay was given the name of "Hawke's Bay" by George Gauld.³

Using *Hawke* as a base ship to which he frequently returned, and assisted upon occasion by her cutter, the surveyor set out in the ship's yawl February 27 to begin his surveys and soundings of lower Mobile Bay. As before, this entailed, first of all, accurate placement and delineation of at least two appropriate and visible points of reference ashore to ensure a corresponding degree of accuracy in the plane perspective of his underwater examinations.⁴ The obvious choices were the east end of Dauphin Island and Mobile Point, to which he ascribed a latitude of 30°16′45″. The magnetic variation was then 4°25′ E.⁵

The rough or field drawing rendered by Gauld in the course of this survey depicts the eastern third of Dauphin Island and all of Little Dauphin Island, which he named "Gillori." Interestingly, he retained much of the well-known eponymous nomenclature of the earlier French explorers and surveyors. On Dauphin Island his rough copy displays several topographic features: the "Bluff" on the extreme east end, the "Thick Trees" of the eastern part, "Marshes," "Path," and an "Avenue cut through," which is marked by two parallel dotted lines from "Hawke's Point" to an area identified on his later fair copy as belonging to Lieutenant Governor Montfort Browne. Browne's property was located on the north shore near the southern terminus

    3. "A Plan of the Bays of Pensacola and Mobile with the Sea Coast and Country adjacent . . . 1768" and untitled rough MS of Mobile Bay [1768]: MODHD, E19/10 5k.

    4. Triangulation, the establishment of reference stations, and the importance of accuracy are discussed in Bowditch, *American Practical Navigator*, pp. 854, 857, 864.

    5. Gauld, "A Plan of the Bays of Pensacola and Mobile with the Sea Coast and Country adjacent . . . 1768," and rough MS of Mobile Bay [1768]. The actual latitude of Mobile Point is 30°13′30″ N. See USC & GS Chart 1266 (23d ed., 1969), *Mobile Bay*.

of the modern Cedar Point–Dauphin Island bridge. The fair copy also depicts a "Guard House" near the gulf beach about two miles from the "Bluff." As the thick woods of the eastern end of the island gave way to low barren land west of present Heron Bayou, Gauld depicted a "Remarkable single tree." To the narrow, tortuous passage between Dauphin and "Gillori" islands he ascribed a depth of one to one and one-half fathoms. Another note identified the small islands extending from the north end of "Gillori" to "Oyster Point," now known as Cedar Point, as "Heron Island, a chain of Dry Keys." Nearby, a faint notation advises that "The Third Island is Pass a l'Herons."[6]

Gauld's written description of these islands and waterways provides further information. Dauphin and the smaller island to the westward, presently named Petit Bois, were formerly one, known as Massacre, and were still so shown on the charts of his day. Christened by Iberville for the "large heap of Human Bones" he found upon landing, it was later called Dauphin Island in honor of the heir to the French crown and to remove the odious connotation of its earlier name. Gauld noted that Father Charlevoix referred indiscriminately to the island as "Massacre" or "Dauphin" and seemed confused as to its size or even whether it was one or two islands. "But it is my business to give an account of the coast as it is at present," Gauld concluded. His familiarity with the early French explorers and their cartography of the Gulf Coast indicates that he had studied their work carefully in preparation for his own.[7]

About ten miles long and not more than one and one-half miles wide at most, Dauphin Island narrowed to a mere slip of land containing nothing but dead trees for about four miles at its western end. The better part of the island was covered with thick pines, which extended close to the water's edge and the bluff on the east end. An "Old French Post on the south side of the Island about 2 miles from that Bluff"—the "Guard House" on Gauld's drawings—and a few old houses on the north side still remained. Near the latter were "large Hillocks of Oyster Shells" covered with dwarf cedar and oak. Gauld noted that there were "many such vestiges of the antient inhabitants in several bays and other places on the coast." They were always found on high banks, where the savages encamped, and could

6. Rough MS of Mobile Bay.
7. Gauld, "A General Description of West Florida, 1769," p. 12.

not, he concluded, have been left by the sea, despite the opinion of many.[8]

A boat might pass between Dauphin and "Gillori" with some difficulty, and between the latter and the mainland was a "Chain of small islands and Oyster Keys" through which there was a passage, four feet in depth, called "Passe a l'Heron"; it provided an inland route for small boats from Mobile Bay to the westward. Gauld noted on the rough and fair copies of his manuscript chart that "There is likewise an inland passage for small boats and canoes from the west side of the Bay of Mobile, thro' what the French call Riviere aux Poules, which falls in opposite to the west end of Dauphin Island and cuts off a considerable space of ground." The first of these passages, now known as Pass aux Hérons, is part of the Intercoastal Waterway, and the second has been dredged to seven feet between East and West Fowl rivers.

After surveying the eastern end of Dauphin Island and Mobile Point, Gauld delineated them on his rough sketch as reference points. He was now able to record accurately all offshore data. Thus Pelican Island and Sand Key, the latter referred to as "Little Pelican Island," and the depths and shoals guarding Mobile were recorded on his manuscripts.

"Great Pelican Island," as Gauld described it, was "just opposite to the Old Fort on the south side of Dauphin Island, distant 1 mile." About one mile long and very narrow, it extended to the southeast in the form of a half-moon, the concave side toward the east end of Dauphin Island. It had neither trees nor bushes but supported "large tufts of grass, like small reeds . . . near the seaside." Between its north end and Dauphin Island was a broad channel of eleven to twelve feet, which opened into the anchorage which he had named "Hawke's Bay." Not only was the spot well protected from most winds, but it afforded a depth of four fathoms and a good holding bottom. Gauld was of the opinion that here was the "remains of the famous Harbour mention'd by Charlevoix to have had formerly a channel of 5 fathoms in it, which was choaked up by a violent hurricane sometime before he visited Louisiana."[9]

Little Pelican — "Sand Key" on the fair copy of his manuscript — lay one league southeast from Pelican Island. A shoal extending from

8. Ibid., p. 13.
9. Ibid.; rough MS of Mobile Bay; Gauld, "A Plan of the Bays of Pensacola and Mobile with the Sea Coast and Country adjacent . . . 1768"; and USC & GS Chart 1266, *Mobile Bay.*

the east end of Dauphin Island to this key protected Hawke's Bay on the east. Continuing more than a mile beyond Sand or Little Pelican Island, it curved to the eastward and joined a large shoal extending south from Mobile Point. Thus was formed Mobile bar with deepest water of fifteen or sixteen feet. Gauld attributed the constant swell and heavy sea during southerly winds to the bold nature of the bar and the surrounding reefs on the seaward side. He thought it imprudent for a vessel drawing more than ten or eleven feet to attempt to cross the bar under such weather conditions, and as his charts showed "Good holding ground without the Bar," discretion was well advised. Although somewhat modified by two hundred years of wind and wave, Pelican Island, Pelican Bay — once Hawke's Bay — and "Little Pelican Island" or "Sand Key" persist to this day.

Of Mobile bar Gauld wrote: "The Mark for going over it in the deepest channel is to bring little Pelican Island well on with the Bluff on the East end of Dauphin Island, bearing about NNW ¾ W and then steer in for the Key in that direction." Once over the bar, the water deepened gradually toward the island, and the channel was not more than a quarter of a mile broad with six or seven fathoms of water. This depth continued all the way around Mobile Point to a "tolerable good anchorage in 6 or 4 fathoms, but it is at best an open Roadstead, the Bay being too large to afford much shelter."

Hawke remained at anchor for twenty-two days while Gauld and his men surveyed the necessary reference stations and performed their hydrographic examination of the bar and entrance to Mobile Bay. On March 8-9, Gauld and the eight men aboard his yawl were driven out to sea by fresh gales and squally weather when their light ground tackle failed to hold the boat against the wind. By March 20 they were back on board, and *Hawke* returned to Pensacola for a stay of one week in order to replenish the crew's provisions. Lieutenant Warburton recorded that he took on "936 pounds of bread, 168 pieces of Beef, 224 pieces Pork, 125 pounds Butter & 8 bushels Pease." The first of April found *Hawke* returned to her task and anchored in lower Mobile Bay one mile north-northeast of Mobile Point, while the cutter, with Gauld and his party, was again off surveying.

Having charted the mouth of Mobile Bay on his previous visit, Gauld directed his attention to its eastern shore. He ascended the Bon Secour River, known now as then by its early French name, for a distance of six miles, at which point he recorded on his rough manuscript that it was "16 ft broad." He accurately delineated Oyster Bay but left it unnamed; as he progressed he noted the sites of "Old Bon Secour," a "Round Clump of Trees," "New Bon Secour," and a "Cowpen." Sounding and accurately outlining Weeks's Bay, he indicated Daniel Ward's place on its south side. He ascended the two streams: the northern still retains its old name, Fish River; the southern, which he called "Le Saut," is now known as Magnolia River. One mile east of Ward's plantation lay those of the French settlers Baptiste and Dupont. Nollie Creek was shown as the "Creek of Good hope." Four miles upstream — the end of Gauld's ascent — was an "Ind'n Camp" and a "Ford 20 ft broad and abt 1½ feet deep rocky bottom where the stream is pretty rapid which they call a waterfall, and that gives name to the Brook vizt. *Le Saut.*" Mullet Point was given the name "Point humide" by Gauld, and "Clear Point," near which he took soundings of eight to nine feet with "hard ground," is now Great Point Clear. Between "Clear Point" and the "Red Cliff" at modern Fairhope lay the property of John Gradinego. The one-hundred-foot elevation now appropriately named Seacliff was indicated as the "highest Red Cliff" on the eastern shore; to the north of it were located Elias Durnford's plantation and a "Yellow Cliff" near modern Daphne. The "Village," which no doubt gave the name to Village Point, lay beyond, as did the plantation of Augustin Rochon. These marked the two ends of the road from Pensacola. Gauld's fair copy of the 1768 survey omitted certain notations on his rough draft but included others, notably the locations of Edmund Rush Wegg's plantation at the southern end of the red cliff at Fairhope and Jeremiah Terry's at the highest yellow cliff at Daphne.

In marked contrast to the eastern shore, virtually the entire western littoral of Mobile Bay for some miles into the interior is low and, in places, marshy. Mon Louis Island, created by the inland passage of the two Fowl rivers, was characterized as "Isle aux Maraguans" on the fair copy of Gauld's manuscript. Hereon he noted a small "Bluff" between Cedar Point and the entrance of East Fowl River. Just south of that river, then known as Rivière aux Poules, he depicted the properties of "Francis" or François and "Mon Louis," once the residence of the Chevalier Montault de Monberaut. "Ward's" place was some three miles from the mouth of the river. The woodlands of the northern part of "Isle aux Maraguans" was marked by "Thick trees," while the southern part was appropriately noted as a "large morass with clay bottom." Gauld depicted only the mouth of Deer River on both his fair and rough drawings, calling it "Buck River" on the former and "Roebuck Riv." on the latter. At the south side of the entrance of "Dog River," the largest branch of which he ascended for a distance of some five miles, he noted the grant held by Pierre Rochon, son of that Augustin who lived across the bay. To complete his toponymy as far north as Mobile, Gauld located McGillivray's plantation and Choctaw Point.[10]

As for the bay itself, Gauld described it as extending about eleven leagues nearly due north from Mobile Point to the town, with a width of three to four leagues except at the lower part. Here a deep bight extended about six leagues eastward from Mobile Point, separated from the sea by a long, narrow peninsula. The soundings on Gauld's manuscript drawings conform in general to the present depths of the bay, which he described as from three to two fathoms two-thirds of the way from Mobile Point to the town and thence only ten to twelve feet, with even less water in places. Large vessels were only able to go within seven miles of town, he advised, "but there is no Danger as the Bottom is soft mud." The inconveniences of navigation notwithstanding, "Mobile having been the Frontier of the French Dominions in Louisiana, always was, and still is a very considerable place."

As Gauld worked his way northward toward the headwaters of the bay, so too did his base vessel, though not without some difficulty. By April 26, *Hawke* had proceeded to a position only three miles southeast of Mobile, where she grounded in seven feet. Running the grapnel, the crew kedged her into deeper water and emptied the barrels of water ballast in order to get nearer to the town. *Hawke* was finally anchored in fifteen feet at the bar of the entrance to Mobile River off Choctaw Point, two miles "off the Town, The Garrison [Fort Charlotte] NW ½ W." In addition to the minor embarrassment of grounding his little vessel, Lieutenant Warburton experienced some disciplinary problems while the surveyor explored the shores of Mobile Bay; he found it necessary to punish two crewmen with twelve lashes each, one for theft, the other for "Mutiny and impertinence."

A line of near-continuous soundings on Gauld's rough manuscript chart indicates that he traversed the headwaters of the bay just south of the marshy islands formed by the Mobile, Spanish, Tensaw, Apa-

10. Gauld, "A General Description of West Florida, 1769," p. 14; rough MS of Mobile Bay; Gauld, "A Plan of the Bays of Pensacola and Mobile with the Sea Coast and Country adjacent . . . 1768"; USC & GS Chart 1266, *Mobile Bay*.

lachee, and Blakely rivers. This same chart also shows that he ascended for some eight miles the first two rivers, both of which then bore their present names. Gauld's fair copy displays complete detailed outlines of the islands and mouths of these rivers, and immediately north of Augustin Rochon's tract he identified "Biminet Bay," now known as Bay Minette Basin.

George Gauld described Mobile as being "pretty regular, of an oblong figure, on the West Bank of the River where it enters the Bay." Its location near the marshes and lagoons at the head of the bay subjected the inhabitants to "Fevers and Agues in the Hot Seasons." He merely noted the presence of Fort Charlotte, known as Fort Condé under the French, and its "neat square of Barracks for the Officers and Soldiers." Gauld observed that several of the richest French inhabitants had left Mobile when the English acquired West Florida, but many still remained in the town and on plantations along the river and on both sides of the bay. Trade with the Indians in skins and furs — the staple business of the community — amounted to twelve to fifteen thousand pounds sterling annually in the London market.[11]

The Mobile River was divided into two principal branches about forty miles above the town, one of which was called the "Tansa." The Mobile River proper was restricted by a bar of seven feet where it entered the bay near the town, but a small branch to the eastward — Spanish River — afforded a least depth of nine or ten feet. Gauld thus indicated the limitations imposed on navigation by the two rivers that form modern Blakeley and Pinto islands.[12]

There is no evidence that Gauld ascended the Mobile River or its branches beyond the few miles embraced by this survey. His subsequent description of the courses of the Alabama, Tombecbé, Coosa, and Tallapoosa rivers could easily have been derived from other sources. Without doubt he was informed of the Tombecbé by his friend Philip Pittman, who had traveled its waters and charted its course in the winter of 1767–68.[13] From the army engineer he would have heard that the Tombecbé was navigable for "sloops and schoon-

11. Gauld, "General Description of West Florida, 1769," pp. 14–15; rough MS of Mobile Bay; Gauld, "A Plan of the Bays of Pensacola and Mobile with Sea Coast and Country adjacent . . . 1768."

12. Gauld, "A General Description of West Florida, 1769," pp. 15–16; USC & GS Chart 1266, *Mobile Bay*.

13. Robert R. Rea, Introduction to Philip Pittman, *The Present State of the European Settlements on the Mississippi*, facsimile ed. (Gainesville, Fla., 1973), pp. xxxvi, l.

ers about 35 leagues above the Town of Mobile" and would have become aware of recent settlers along its banks who found the soil "to answer beyond expectations" and praised the diversity of its timber. Gauld noted that the chief settlements of the Upper Creek Indians were along the "Coussa" and "Talypouse" and their branches, and that the "French Fort at Alibama," Fort Toulouse, had been evacuated in 1763 and never garrisoned by the British. The fort at Tombecbé, Fort York, beyond which lay the country of the "Chicasaws," had been occupied by the British but was abandoned in 1767, when Pittman visited it, "by order of the Commanding Officer of the Troops at Pensacola."[14]

By the first day of May 1768, Gauld had completed his survey of Mobile Bay and its environs and returned to *Hawke*. Five days later the ship got under way and proceeded slowly down the bay, dragging through the mud in a depth of six feet for a mile before reaching deeper water. On May 9, *Hawke* and the survey party were back in Pensacola, but they tarried only three days in order to replenish their provisions and secure a new anchor cable from H. M. S. *Adventure* before sailing westward to resume Gauld's surveys.

The next stage of this often-interrupted phase of Gauld's work concentrated upon the seacoast beyond the great bays of Mobile and Pensacola. Before him lay the task of completing the remainder of the shoreline, bays, and bayous of the present states of Alabama and Mississippi and examining the Pascagoula and Pearl rivers and the lakes and bayous touching the Isle of Orleans and Spanish Louisiana. Along with this he must examine and chart the offshore islands that formed the relatively shallow inland waterway now known as Mississippi Sound and the passageways between these islands.

The survey ship got under way shortly after dawn May 12, 1768, but encountering contrary winds and a strong head current, which brought with it the danger of many large drifting logs, Lieutenant Warburton decided to anchor off Pensacola bar and await more favorable conditions. It was therefore three more days before *Sir Edward Hawke* moored near Dauphin Island in the bay that now bore the ship's name.[15]

Fresh gales with squalls and rain kept Gauld's party on board the base vessel for the next six days. The seventh day brought clear skies

14. Gauld, "General Description of West Florida, 1769," p. 16. On the British occupation see Robert R. Rea, "The Trouble at Tombeckby," *Alabama Review* 21 (1968): 21–39.
15. That is, Pelican Bay. Captain's log, *Sir Edward Hawke:* ADM 51/4342.

Surveys to the West, 1768

and moderate winds, allowing Gauld to set out in the cutter with a crew of sixteen men to resume his work. Taking up where he had left off, near the eastern end of Mississippi Sound, he surveyed and drafted on his field sheets the coastline as far as the west arm of the river known then as now by the name Pascagoula. Gauld's working drawings indicate that he ascended the east branch of the river some eight miles to "Pt. de l'auvergue" and the west branch about an equal distance. The coastal examination took him into the "Bay de Baros" (Portersville Bay, into which Bayou Le Batre flows), past "Pt. aux Herbes," along the shoreline of a large bay unnamed by him but now known as Grand Bay, around and over "Grande Batture or The Great Bank," and into "L'ance de Grande Bature," now known as Pointe aux Chênes Bay. From this bay Gauld rounded the Pointe aux Chênes, briefly examined part of the "Bay de Coussot" (Bayou Casotte), passed Pascagoula Bluff, and finally made his way into the west arm of the river and proceeded through a navigable cutoff to the east branch and thence to Hugo Krebs's plantation. Here the cutter anchored while Gauld enjoyed the hospitality of the most prominent planter on the coast. Krebs's plantation is now part of the city of Pascagoula, but his name is perpetuated in the nearby cove known as Krebs Lake.[16]

In describing the Pascagoula River, Gauld remarked that it emptied "through several mouths" and between the easternmost and westernmost, a distance of about four miles, was a continuous bed of oyster shells showing only sufficient depths for a boat. The sole navigable channel was in the westernmost branch, which had a depth of four feet. About twenty miles above its mouth this large river divided into two branches which descended southward about five or six miles apart with "nothing but low marshes, lagoons and intersections from one to the other" for several miles before they reached the sea. Inside the entrances, both branches deepened to three to six fathoms and were reported "to be navigable for upwards of 50 leagues to the heart of the Choctaw Nation." The soil along the river improved as one ascended; there were good plantations on the east side of the river, but

---

16. Ibid. Gauld MS field sheets of Mississippi Sound from "Bay de Baros" [Porterville Bay, Ala.] to "The Lake Maurepas" [1768]: MODHD, E19/13, 4a, cited hereafter as "Field Sheets of Mississippi Sound and Lakes, 1768–1769"; USC & GS Chart 1267 (14th ed., 1968), *Mississippi Sound and Approaches*. The alphabetical labeling and sequential numbering on Gauld's field sheets, when considered with the movements of the base vessel as reflected in the captain's journal, indicate the geographical order of this phase of his surveys.

the Indians, particularly the Choctaws, molested the inhabitants and killed their cattle. Firmly attached to the French during their possession of Louisiana, Gauld thought it would be some time before the Choctaw would be thoroughly reconciled to the English. The mainland coast between "L'Ance de la Grande Bature" and Mobile Bay was swampy, with a clay bottom for two or three miles back. It gave way to sandy or gravelly soil supporting pines, live oak, and hickory. Several miles from the coast the soil improved to a degree suitable for cultivation and pasturage, and cattle raising might "turn out advantageously" if pursued by industrious inhabitants.[17]

From the mainland, Gauld proceeded to Dauphin Island, where he completed his survey and delineation of its western end. Working farther westward he examined and charted "Massacre Island" (Petit Bois Island) and the eastern part of Horn Island. At this juncture he crossed the sound to Round Island and thence to the west arm of the Pascagoula, which he surveyed and sounded for some eight miles. Returning to the river's mouth, he examined the coastline to the west as far as the entrance to the Bay of Biloxi, noting en route the mouth of Graveline Bayou and the broad curve of Bellefontaine Point.[18]

Having established these landmarks on his manuscript, Gauld conducted soundings offshore of the island, extending these notations as far west as his inshore observation of depths and bottom characteristics. He noted that the passage between "Massacre" and Dauphin Island had "Breakers all the way." Between "Massacre" and Horn Island, two miles west, a shoal extended for about a mile and a half, leaving an eleven-foot channel at the western tip of "Massacre." Anchorage in four fathoms was available to vessels able to cross this bar. Gauld noted that the "common charts" took no notice of this island. About nine miles long and very narrow, it took its name from a grove of trees midway of its length which Gauld labeled "Petit Bois de Massacre." This "small wood" was all the more remarkable since not another tree stood on the island. The acuity of George Gauld's observations may be judged by the fact that the dredged channel for ships utilizing the modern facilities of the port of Pascagoula now runs through the natural deep water hard by the west end of Petit Bois Island which Gauld first charted.

17. Gauld, "A General Description of West Florida, 1769," pp. 11–12.
18. Gauld, "Field Sheets of Mississippi Sound and Lakes, 1768–1769"; USC & GS Chart 1267, *Mississippi Sound and Approaches*.

Surveys to the West, 1768

Between "Massacre" and the mainland, the sound was about ten miles wide with depths of two to three fathoms. An exception was noted in the large shoal called "La Grande Bature" extending from the mainland about one league. Having only two or three feet of water or less, this shoal lay at the opening of a broad bay known as "L'Ance de la Grande Bature" or "tongue of the great shoal."

En route from Horn Island to the mainland, Gauld examined and placed on his manuscript the small island still known as Round Island. Covered with tall trees, Round Island and the bluff at the east entrance to the Pascagoula River were both clearly discernible over the eastern end of Horn Island. An "old hut" and "an old house" on Horn Island, and an "old house" on Round Island, were depicted on Gauld's field sheet, indicating that these islands had been occupied at one time by civilized inhabitants.[19]

Gauld set out on the first day of his survey of Mississippi Sound, its offshore islands, and the opposing mainland on May 23 and remained away from *Hawke* one week. He was met with bad news upon his return. Although the base vessel had been heeled and her bottom payed with pitch and turpentine in early January, she was found to be leaking badly. Not only had the worm taken its toll, but she had been hard and fast aground on several occasions despite the precaution of sounding ahead with a small boat. Lieutenant Warburton insisted that he must return to Pensacola for repairs, and since Gauld recognized the importance of utilizing to the fullest the relatively good weather of spring and summer, he and Warburton decided that they would provision the cutter for thirty days and that Gauld would remain behind to continue his surveys with a crew reduced in number to twelve men. On June 3 they parted.[20]

Now independent of a base ship, Gauld proceeded to examine and chart the Bay of Biloxi and the coastline westward to a position near St. Louis Bay. In like manner, he surveyed the west end of Horn Island and all of Ship Island with their intervening passes, channels, and shoals. To complete this section, the bottom characteristics of Mississippi Sound as far as Cat Island were taken and recorded on his field sheets.[21]

Commenting upon this stretch of coast Gauld observed that "Just

19. Ibid.; Gauld, "A General Description of West Florida, 1769," pp. 11–12.
20. Captain's log, *Sir Edward Hawke:* ADM 51/4342; extract of letter, Warburton to Parry, November 12, 1768: ADM 1/238.
21. Gauld, "Field Sheets of Mississippi Sound and Lakes, 1768–1769."

opposite to Ship Island on the Main Land is situated Old Biloxi on a small Bay of the same name, behind L'Isle au Chevreuil, or Buck Island." Biloxi had been the first French establishment in Louisiana, but finding it unsuitable for the provincial capital, the early settlers soon abandoned it, and only a few of their descendants remained there in 1768. Their chief means of livelihood was the raising of cattle and the production of pitch and tar, but even in this they were troubled by the savages. Gauld found that the Bay of Biloxi, as well as the adjacent coast, was very shoal and fit only for small craft. Two small rivers or creeks flowed into the bay, but neither was of any consequence.[22]

Having surveyed and sketched Horn Island and Ship Island on his manuscript, Gauld provided a detailed description of each. They were separated by a distance of about six miles, and a small key, Dog Island, lay between them. Apart from a narrow channel of five or six fathoms hard by the west end of Horn Island, the passage between the islands was closed by a continuous shoal, dry in places at low water. The channel was restricted by a bar of seven feet and further guarded by an easterly sweep of the shoal extending offshore. Great care had to be exercised in entering the channel from westward or southward by anyone attempting to use the anchorage within the west end of Horn Island. An approach from eastward was relatively safe, as deep water prevailed all along the seaward side of Horn Island and around its west end into the anchorage. Close inshore along Horn Island, the bottom was soft mud mixed with sand, but farther offshore, in eight or nine fathoms, the mud gave way to brown sand with black specks and some broken shells. Horn Island was some sixteen miles long and not more than a half mile wide, oriented nearly east and west. Its vegetation was varied in density if not in kind: straggling trees were noted on the west end; a rather thick growth flourished in the middle; but only skeletons stood toward the eastern extremity, which was devoid of living trees, a jumble of sandy hummocks. Depths of two to three fathoms prevailed almost all the way from the island to the mainland.[23]

With the possible exception of Dauphin, the most important of this chain of offshore islands was Ship Island. Nine miles long and some four leagues from the mainland, within its western end it afforded a three- to four-fathom anchorage which was well protected

22. Gauld, "A General Description of West Florida, 1769," p. 9.
23. Ibid., pp. 9–10; Gauld, "Field Sheets of Mississippi Sound and Lakes, 1768–1769."

from offshore winds in summer but exposed to the northers of the winter months. Gauld provided explicit sailing directions for vessels approaching from the east and crossing the twenty-two-foot bar into the anchorage. Similarly, he indicated the best positions for frigates and smaller vessels. Presently known as Ship Island Pass, the dredged channel to modern Gulfport, Mississippi, passes through the deep-water route described by George Gauld. Of great importance to the eighteenth-century seafarer was the availability of fresh water and firewood on Ship Island, and nearly as important was a herd of 150 to 200 cattle from which fresh beef was supplied to His Majesty's and other ships by the family then residing on the island. In season, turtle and fresh fish were to be had in abundance.[24]

Gauld worked steadily to the westward and at length found himself taking soundings some two miles offshore of the south side of Cat Island. About six miles long and three wide, Cat Island had a channel of nine or ten feet, its south side separating it from the marshy islands bordering the Spanish Isle of Orleans. Separating the north end of Cat Island from the mainland of British West Florida were three leagues of open water interspersed with a "great many shoals and oyster banks" between which several passes might be found deep enough for such vessels as were able to enter the lakes.[25]

Off Cat Island, George Gauld experienced the first of several chilling and near-tragic experiences with the cutter. The little vessel had been leaking badly for some time when suddenly water began to come in so fast that it was only with great difficulty that Gauld and his men were able to keep her from sinking. Fortunately, a fresh southwest wind enabled them to make a run for the shore, and they beached the cutter just before she filled. Upon hauling her out of the water they found that "the plank next the keel had started upwards of three feet, the nails being given away."

Repairing the damage as best they could, Gauld and his men were soon back at their work — only to find themselves in difficulties once again. Caught in a southeast gale among the keys near Cat Island, and being unfamiliar with the shoals between them and Bay St. Louis, they had no choice but to ride out the high seas and heavy winds near one of the keys. They were compelled to remain there "two nights and one day at a Grapnel, the greatest part of the time with the whole length of the Grapnel Rope, and if either of them had

24. Gauld, "A General Description of West Florida, 1769," pp. 8–9.
25. Ibid., p. 4.

given away, there was a thousand chances against us." After this experience Gauld and his men had had enough of narrow escapes, and they were nearly out of provisions. When the weather moderated they got under way and beat back to Cat Island, and on the following day they set out for Horn Island, their rendezvous with the base vessel.[26]

As for *Hawke*, her repairs had been extensive. With the assistance of carpenters from H. M. S. *Druid*, she had acquired a new mainmast and had had her bottom tarred. On July 7 she sailed from Pensacola, and the following day, off Mobile Point, Lieutenant Warburton began making the prearranged signal, one gun fired every two hours, as he proceeded slowly westward. The next day, and many signals later, Gauld and his men in the cutter came alongside *Hawke* near Horn Island Pass. Together the two vessels proceeded directly to Ship Island.[27]

Having consolidated his field notes and recorded his examinations during *Hawke's* absence, Gauld took provisions for one month aboard the cutter and set out to survey the Chandeleurs and the Breton Islands farther to the south; he then intended to sweep as much of the coast as possible toward the entrance to Lake Pontchartrain. Less than a fortnight elapsed before the cutter again showed serious leaks, but by unshipping the foremast Gauld and his crew were able to ease her in the seaway and keep her from filling. On the night of August 7 the watch aboard *Sir Edward Hawke* "heard the report of several guns in the offing"; replying with a series of shots from his swivel gun, Lieutenant Warburton guided Gauld and the cutter back to the safety of Ship Island and their vessel. Hauling the cutter out of the water, they discovered that not only had the earlier leak recurred but others had also developed. To make bad matters worse, several broken timbers were discovered. As Gauld later observed, the cutter "was always extremely weak, especially forward."

It was obvious to Gauld that without thorough repairs the cutter's usefulness as a survey boat had come to an end, and as good summer weather was precious, it was decided to continue with the ship's yawl, as long as provisions lasted, rather than return to Pensacola. The cutter was put out to anchor, but even then her troubles were not over. During the night of August 11, fresh gales and squally weather parted her cable and she blew ashore. Lieutenant Warbur-

26. Extract of letter, Gauld to Parry, Pensacola, November 19, 1768: ADM 1/238.
27. Captain's log, *Sir Edward Hawke:* ADM 51/4342.

Surveys to the West, 1768

ton was able to refloat her the next day, but *Hawke*, too, was in need of repairs, in spite of her recent careening and caulking and the new mainmast.[28]

George Gauld was probably able to complete his examination of the chain of islands extending southward from the Mississippi Sound toward the Mississippi River delta before the cutter failed him the second time. He described the Chandeleurs as being five in number and extending nearly south by west for nine or ten leagues. Originally called the Myrtle Islands by Dr. David Cox, an adventurer of the era of King William III who also discovered the adjacent Nassau Road, their name was changed to the Chandeleurs by the French who utilized the wax of the abundant myrtle bushes to make their candles. Fresh water could be obtained by digging or from a well at an old hut near the north end of the islands, but only driftwood was available for fuel. Ten or eleven miles southward, across a continuous line of breakers, lay the Isle aux Grand Gosiers, or Great Pelican Island, and to the westward of this some four miles was the group known as the Bretons. All were very low and treeless. The low land and intervening shoal between the southernmost of the Chandeleurs and Grand Gosier, from latitudes 29°42' to 29°32' N, made this a dangerous part of the coast. Behind these outlying islands lay the "chain of low marshy islands and lagoons bordering the Island of Orleans."[29]

Three good anchorages lay within the protection of these offshore islands, two for smaller vessels and the third capable of use by the largest vessels that could come to the coast of Florida, according to Gauld. Within the Isle au Grand Gosier a smaller vessel might "ride in safety from easterly winds in 4 fathoms, soft ground," and Gauld described the anchorage and provided specific directions for gaining access. An anchorage of sixteen to seventeen feet lay within the southwest point of the Isle au Breton. For seamen frequenting the Mississippi, a knowledge of these anchorages was of the utmost importance. Within the north end of the largest of the Chandeleurs, five leagues south of Ship Island, lay Nassau Road, so named by Dr. Cox for the English ruling house at the time he discovered it. Easily accessible from the sea and free of any bar, the only precaution to be observed in entering the anchorage was to stay at least

---

28. Ibid.; extract of letter, Gauld to Parry, November 19, 1768: ADM 1/238.
29. Gauld, "A General Description of West Florida, 1769," p. 2; Gauld, "Field Sheets of Mississippi Sound and Lakes, 1768–1769."

three-quarters of a mile from the inside of the island. It afforded good shelter from easterly and southerly winds and had a depth of four fathoms with good holding ground. This he judged to be "one of the best [anchorages] for large ships on the whole Coast of Florida," and because of the frequency of easterly winds he considered it essential "for his Majesty's ships to be well acquainted with this place."[30]

After a few days spent completing nearby surveys with the yawl and a six-man crew, Gauld returned on board *Hawke*. Proceeding into Cat Island Channel, the ship was anchored in the protected waters of Spit Cove, formed by two arms of the island. Remaining only briefly, *Hawke* sailed farther westward, anchoring, as was customary in unfamiliar waters, during the hours of darkness. On August 31 Warburton ran aground in seven feet of water on the bar as he attempted to enter Lake Pontchartrain by the way of the Rigolets. After three hours' effort, *Hawke* was freed and retraced her course to the nearby Isles de Malheureux, where she anchored in two fathoms. No further attempts were made to take the ship into the lakes. Gauld later observed that had the schooner drawn only six or seven feet, "she would [have been] of much greater service."[31]

While Lieutenant Warburton observed the considerable coastal shipping between New Orleans and Mobile, the surveyor continued his work in the ship's yawl. Toward the entrance to Lake Pontchartrain, Gauld encountered the small compact Bay of St. Louis, which had only seven or eight feet of water. The land was "pretty good, especially for pasturage," Gauld wrote, but the several settlers had been forced to leave in 1767 because of marauding Choctaws who killed their cattle. About three leagues farther on were "Les Isles aux Malheureux and some others," two of which he identified on his field sheets as "St. Joseph's Island" and "Heron Island" (proper names still to be found on the map). There was deep water and good anchorage among them "for such craft as go that way." From thence to the Rigolets was about two leagues. The principal channel into Lake Pontchartrain, this waterway was approximately three leagues in length and about three or four hundred yards broad with marshes on each side. Like the passage of Chef Menteur to the south, the Rigolets had a limit in depth of seven feet.

30. Gauld, "A General Description of West Florida, 1769," pp. 2–3, 8.
31. Extract of letter, Gauld to Parry, November 19, 1768: ADM 1/238; Captain's log, *Sir Edward Hawke*, August 18–31, 1768: ADM 51/4342.

Surveys to the West, 1768

Gauld's field sheets clearly indicate that he eventually surveyed, sounded, and charted Lakes Borgne, Pontchartrain and Maurepas, their connecting and tributary waters including the Rigolets and Chef Menteur, the communications between Lakes Borgne and Pontchartrain, and Pass Manchac between Pontchartrain and Lake Maurepas. Among the tributaries he examined were the multi-branched Pearl River, Bonfouca Creek, La Combe River, the modern Tickfaw and Natalbany rivers, the River Iberville or Manchac, and, finally, the Houmas River. His farthest advance westward on this expedition, however, took him through the Rigolets into Lake Pontchartrain and thence into Bayou St. Jean, characterized by Gauld as "the back door to New Orleans"—a portal he was quick to open. Gauld depicted New Orleans as a rectangle, nine blocks along the riverfront and four deep. Connecting the city with "Bayou Villax" was a road, about one mile in length, on which he identified an old brick kiln, and where this road crossed a small branch there was a "watering place." On the west side of the entrance of Bayou St. Jean into Lake Pontchartrain was a guardhouse, and opposite it were some fishermen's huts.

Remaining on survey some twenty-five days, Gauld was compelled to return to *Hawke* at weekly intervals because of the yawl's inability to carry provisions for him and his crew for a longer period, and apart from visiting New Orleans it is doubtful that he was able to chart much more than the Rigolets and the south shore of Lake Pontchartrain at this time. Details of his surveying of the lakes and rivers to the northward are obscure. Although he remarked in November 1768 that he had been "hard at work making out a fair copy of the survey" as far as he had gone, this fair copy has not been found. Consequently, the full extent of his work on the lakes for this year can only be surmised from the time he remained there and his vague statement that he stayed a "few days" in New Orleans.

By September 25, George Gauld and his crew were aboard *Hawke* and standing to the eastward. Their return was attended with much difficulty. It was eight days and five groundings later—the last requiring the removal of his iron ballast in order to free the ship—before Lieutenant Warburton anchored in the now familiar cove within the west end of Ship Island. Here the crew "stayed the masts and set the rigging" and replenished their supply of firewood and fresh water. On October 5 all was in readiness and *Hawke* set sail for

Dog Island, where Warburton worked back and forth along the offshore islands, presumably to allow Gauld to complete his soundings. The cutter, meanwhile, was taking water at such a rate that two men could scarcely keep her afloat. Eight days later, October 13, *Hawke* anchored in Pensacola Bay, and the ill-fated cutter was immediately sent for repairs to H. M. S. *Renown,* which, along with *Druid* and several merchant ships, was present in the harbor. Return to Pensacola brought an end to Gauld's surveying for 1768, but his work was far from over. Almost a month was required to incorporate his rough drafts into the fair copy which he transmitted with a covering letter to Admiral Parry on November 19. Proud of his recent accomplishments, Gauld flattered himself that the admiral would note the increased productivity of the current year compared to that of previous seasons, an improvement which he credited to Parry's having assigned a ship exclusively for his work. Although Gauld noted that *Hawke* drew somewhat more water than he considered desirable, he took great satisfaction in the work he had completed with the help of the vessel.

His health had been good throughout the summer, and he had been unsparing of his naturally strong constitution and energies in pursuing his work. He acknowledged, however, that he was "thoroughly tired," and not having allowed himself a holiday since the middle of February, he would welcome a respite from his duties. "Even the few days we were at Orleans were not thrown away, as I took an opportunity of sketching the Town and Environs of it," Gauld remarked in closing.[32]

Although Gauld and his servant were enrolled on the muster table of *Hawke* until April 26, 1769, the surveyor made no further use of the ship. In addition to Gauld's complaint that her draft rendered her unsuitable for survey purposes, Lieutenant Warburton confirmed that the invasion of the worm had left her bottom leaky and that her timbers were badly strained from having gone aground so many times. On the other hand, since Gauld's work for the next year would be conducted "mostly on an open and inhospitable coast," Warburton thought it inadvisable that the surveyor should be left unattended with the cutter while *Hawke* was being refitted. The past season's ex-

32. Extract of letter, Gauld to Parry, November 19, 1768: ADM 1/238; Gauld, "Field Sheets of Mississippi Sound and Lakes, 1768–1769"; USC & GS Charts 1268 (11th ed., 1969), *Lake Borgne and Approaches,* 1269 (15th ed., 1969), *Lakes Pontchartrain and Maurepas.*

Surveys to the West, 1768

perience had shown the importance of maintaining contact between the surveyor and his base ship, but it was time for Gauld and *Hawke* to part.[33] From survey work *Hawke* and her crew were assigned to building a wharf near the careening place on the south shore of the peninsula opposite Santa Rosa Island. This finished, the vessel was given a complete overhaul and sent on a voyage to Campeche, Mexico, in company with the brig *Active,* and the two vessels did not return to Pensacola until April 15, 1769.

33. Extract from Lieutenant Warburton's letter to Parry, November 12, 1768: ADM 1/238.

CHAPTER VIII

Service Ashore and at Sea, 1769–1771

By the winter of 1768–69, George Gauld had become a prominent figure among the residents of Pensacola. In spite of long absences during the surveying season, he was a property owner and householder; he had indirectly served both the governor and the commander of the military forces in West Florida, and he was a permanent fixture — practically the only one — in the ever changing naval establishment at Pensacola. He had seen the provincial capital emerge from its crude Spanish origins to become a promising center of commercial activity. He had also seen West Florida's dreams of mercantile wealth falter in the face of restrictive British regulations and the superior advantages enjoyed by New Orleans in the Indian trade. Along with the rest of the colony Gauld had most recently experienced the shock of learning that the British regiments stationed in West Florida were to be withdrawn, a threat to the security of the colony but even more a blow to its economy, for the army was the most reliable source of business and ready money. After returning to Pensacola in October 1768, Gauld would have seen transport after transport sailing from the harbor as General Haldimand embarked his troops for St. Augustine and other distant ports. The very life of Pensacola seemed at stake, and the community rallied its strength in a fight for survival.

Led by Lieutenant Governor Montfort Browne, in the absence of the appointed royal governor, Pensacola showered protests and petitions upon the imperial authorities in London. Merchants bewailed the loss of their investments; the colonial assembly warned of dangers from hostile Indians and belatedly undertook to pass legislation required by the suddenly changed conditions. In the midst of that work, the term for which the assembly had been elected came to an end, and it was dissolved. Concerned to see the task completed, Lieutenant Governor Browne and his council decided, on December 15, 1768, to order elections for a new assembly to be held early in January 1769. Among the eight representatives who would be returned to the assembly by Pensacola was the naval surveyor George Gauld.[1]

The Scots mathematician and cartographer was not a partisan politician. In a society where place and influence were sometimes dubiously but always vigorously used to advance men's wealth and profits, where smuggling and illicit trading were almost respectable callings, there is no evidence to suggest that George Gauld looked beyond the proper horizons of his professional position. He was, however, one of the elite and privileged class, a highly educated man, possessed of a strong sense of duty. No doubt any feeling of obligation to the welfare of the colony on his part was nurtured by his good friend Dr. John Lorimer, an army surgeon who shared scientific interests with Gauld and who had served in every session of the West Florida General Assembly.[2] By the middle of December the surveyor had completed fair copies of his latest surveys and had dispatched them to Admiral Parry; the season was unfit for further exploration, and West Florida was in need of the services of good men. George Gauld qualified, as a property owner, for a seat in the Commons House, and in all probability he had little to do but offer himself to the electorate of Pensacola in order to win a seat in the Third General Assembly, which convened on Wednesday, January 25, 1769.

Gauld was the only freshman member of the Pensacola delegation in the Commons and one of the few members of the house without previous legislative experience. Both his personal reputation and his freedom from involvement in earlier quarrels between the Commons and Lieutenant Governor Montfort Browne recommended him for a prominent role in the assembly's proceedings. At the initial meeting of the house on January 25, Gauld was one of two members assigned to attend the lieutenant governor and advise him of the house's choice of a speaker; with John Lorimer he helped to prepare and presented to the house "an Humble Address of Thanks" to Browne for His Honor's speech at the beginning of the session; and he and Lorimer, among others, were appointed to the Committee of Privileges and Elections and the Grand Committee for Grievances, which also served as a corresponding committee with the provincial agent and was responsible for preparing the official record of the house. The

1. Robert R. Rea and Milo B. Howard, *The Minutes, Journals, and Acts of the General Assembly of British West Florida*, pp. 142–43.
2. Robert R. Rea and Jack D. L. Holmes, "Dr. John Lorimer and the Natural Sciences in British West Florida," *Southern Humanities Review* 4 (1970): 363–72.

Commons' address to Browne, a warm and cooperative response to his opening speech, was well received by the lieutenant governor; the session seemed off to a good start, but mutual good will proved to be as short lived as the session itself.[3]

As chairman of the Committee of Privileges and Elections, Gauld was quickly pushed into the cauldron of political squabbling. His first unpleasant task was to move the house to arrest the lieutenant governor's private secretary, who was accused of insulting the committee. The offender was brought before the house, required to submit a written apology, and fined. More controversial and time-consuming was the investigation by Gauld's committee of the disputed return of the members for Campbell Town. The Commons ultimately declared their election irregular, found the provost marshal guilty of improprieties, and ordered the election return altered. A more dangerous contest soon developed over the respective prerogatives of the governor and the house. David Ross having been returned for both Mobile and Pensacola and choosing to sit for the latter, the house sought to fill the resulting vacancy by issuing an election writ through its speaker. Montfort Browne denounced the action as an encroachment upon the governor's prerogative. The Committee of Privileges and Elections was ordered by the house to search for precedents supporting its claim, and the task fell to George Gauld. As they "found several precedents" in their favor, Gauld and the committee were directed to prepare another "humble address" to His Honor asserting the rights of the speaker and the house. Lieutenant Governor Browne curtly rejected the house's evidence and arguments, but he sought to avoid a confrontation by adjourning the assembly from February 2 until March 10, when he hoped the problem might be resolved by Governor John Eliot, who was expected to arrive any time. The house responded with a fiery set of resolutions denouncing Browne's action as prejudicial to "the welfare of this Colony." George Gauld's participation in every session of this General Assembly brought him to political prominence and placed him on the side of the Commons against the unpopular lieutenant governor. If the language he used was respectful, it was also the language that had led to limited monarchy in England and would lead to the rejection of all monarchy in America.

The naval surveyor took advantage of the prorogation of the Gen-

---

3. On the Third General Assembly, see Rea and Howard, *General Assembly*, pp. 162–73.

eral Assembly to return to his work. Apparently making use of the cutter, Gauld sailed east along the coast in February, and in March he reached a point "ninety miles beyond Apalache." Gauld's party entered the Suwannee River, and on April 1, he camped about fifteen miles above its mouth. He seems to have explored some twenty miles farther up the river before returning to Pensacola some time late in April. There he found both a new governor and a new naval vessel to which he would be attached.[4]

The universal hope that Governor John Eliot would bring order to West Florida politics was destined to disappointment. A twenty-six-year-old navy captain with exceptionally useful political connections, Eliot, like George Gauld, had served under Arthur Forrest during the Seven Years War. Appointed governor of West Florida in 1767, his departure for America had been delayed by changes in the imperial administrative system, and he did not reach Pensacola until April 2, 1769. He was received with enthusiasm, and he attacked the multitudinous problems of his government with sensible firmness and moderation. The assembly, which had been further prorogued pending his arrival, was soon dissolved and new elections were ordered. Responsible businessmen and politicians were added to the council, and among the appointments that a new governor had to make, Eliot named George Gauld to the rota of justices of the peace for the district of Pensacola. Bereft of military support, threatened by Indian disturbances at Mobile, and unsure what might result from the revolt of the French population of New Orleans against its first Spanish governor, John Eliot husbanded the slender naval defenses of West Florida. When Gauld wished to embark upon another voyage in April, the governor insisted that he delay his departure. Consequently the surveyor was in Pensacola on May 2, when young Eliot, apparently suffering from a brain tumor, put an end to his life and left the colony once more in the hands of Montfort Browne. George Gauld was among the fifty-three distinguished Pensacolans whose signatures attested their sincere grief at John Eliot's "sudden death" and lamented that it had blasted their fondest hopes for the colony's future.[5]

4. Capt. William Philipps to Admiralty, May 14, 1769: ADM 1/2301; *An Accurate Survey of the Coast of West Florida* (London, 1803).

5. Robert R. Rea, "John Eliot, Second Governor of British West Florida," *Alabama Review* 30 (1977): 243–65, and "The Naval Career of John Eliot, Governor of West Florida," *Florida Historical Quarterly* 58 (1979): 451–67.

Before his death, the late governor had provided for the election of a new assembly; it met, according to his intent, on May 22, 1769, and George Gauld sat a second time as a member for Pensacola. Relations between the Commons and Lieutenant Governor Browne were inevitably tense, for eight of the new members had formally expressed their "most disagreeable apprehension for his past conduct" and their horror at beholding "his second approach." For his part, Browne identified his enemies as "a Scotch party," and though he did not name George Gauld, he may well have numbered the Banffshireman among them.[6]

The minutes of the first session of the Fourth General Assembly of West Florida do not record attendance, so it is impossible to determine Gauld's participation except that he was named to various committees, including the Privileges and Elections Committee. It is most likely that he attended faithfully, and as a member of the committee for preparing and bringing in bills he would have been regularly involved in the legislative process. So heavy was this task that after meeting six days a week for four weeks, the house requested a week's adjournment, June 15, in order to put their paperwork in order. The session was concluded on June 29, at which time the lieutenant governor gave his assent to six bills and prorogued the assembly until August 7. Unhappily, relations between Browne and the Commons had not improved, and the Commons had failed to provide taxes for the colony's many needs. What might have produced a financial crisis elsewhere created no more than an impasse in West Florida, however, for the provincial administration was supported by the imperial government in Westminister.[7] George Gauld, who likewise drew his salary from the crown, must have sighed with relief at the end of a hectic session and the opportunity to put such matters behind him and return to the sea and sands that waited to be surveyed.

On May 1, Gauld and his servant, Peter Quash, had been transferred from the rolls of *Sir Edward Hawke* to the muster of Captain George Talbot's sloop *Jamaica* by order of Captain William Philipps. Commander of H. M. S. *Tryal*, in which Governor Eliot had come to

---

6. "The Humble Remonstrance of the Inhabitants of the Town of Pensacola," May 2, 1769: C.O. 5/627. The origins of the political label seem to go back to the hostility between Browne and Governor Johnstone, a Scot. Having enjoyed Johnstone's favor, Gauld might well have fallen into Browne's categorization, and his association with the opposition in Commons in 1769 would have marked him among Browne's enemies.

7. Rea and Howard, *General Assembly*, pp. 174–76, 186–205.

Pensacola, Philipps remained on the West Florida station for some time as the senior naval officer specifically responsible for the colony's defenses by sea. *Jamaica* was in Pensacola harbor in mid-April, and at that time Gauld had intended to embark for the Chandeleurs, but Eliot had forbidden the surveyor to sail; Lieutenant Governor Browne had asked that he delay his departure after Eliot's death in May, and his duties in the Commons House kept Gauld ashore until the middle of June.[8]

This politically tumultuous season was not devoid of other interesting developments for Pensacola's little band of gentlemen-scientists. The Anglo-American community of astronomers was preparing to observe a rare phenomenon, the transit of Venus across the face of the sun on June 3, and at the capital of West Florida, John Lorimer and George Gauld planned to combine their talents and equipment to make the most of this celestial event. Observation of the transit would help to establish the solar parallax and the distance to the sun. Gauld's professional interest in such matters had been stimulated by the solar eclipse of January 19, 1768, and two years earlier Dr. Lorimer had determined the longitude of Pensacola by calculations based upon the eclipses of the satellites of Jupiter. Both men were clearly excited by the prospect of participating in one of the great scientific moments of the eighteenth century. Unfortunately, in April, Dr. Lorimer was ordered to take up residence in Mobile, where his medical skills were desperately needed, and the two friends were forced to conduct separate observations. For Lorimer, removal to Mobile was catastrophic; the weather there was so cloudy and rainy that he was only able to secure a brief glimpse of the transit of Venus "through some flying clouds." At Pensacola, George Gauld enjoyed fair weather and "had a very complete observation and . . . took the opportunity of the distinct horison which that commodious Bay afforded for correcting his time piece by some observations with a good reflecting quadrant, as also by several altitudes taken with a large theodolite during the time of the transit."[9]

8. Philipps to Admiralty, May 14, 1769; Browne to Philipps, May 13, 1769: ADM 1/2301.
9. John Lorimer to ?, Mobile, June 24, 1769: Letters and Papers, Royal Society, London, V, 49–75, microfilm copy, American Philosophical Society, History of Science Film No. 1, Reel 8: 4344–47; Maxine Turner, "Dr. John Lorimer's Observation of the Transit of Venus at Mobile in 1769," *Georgia Journal of Science* 35 (1977): 136–39. Robert R. Rea is indebted to Prof. Turner for bringing this episode to his attention. See also Harry Woolf, *The Transits of Venus* (Princeton, 1959).

Gauld sent his findings to Lorimer and left it to the surgeon to draft their report to the Royal Society. The surveyor's ship was ready to sail, and he returned to his coastal investigations on June 17, without waiting for the final session of the General Assembly. Captain Talbot carried the surveyor to Ship Island, and from that safe anchorage Gauld began the summer's work, once more utilizing the twelve-oared cutter with a crew of two petty officers and ten men from *Jamaica*.[10]

George Gauld had repeatedly urged upon his superior, the commander in chief at Jamaica, the great desirability of having a shallow-draft vessel for surveying the coastal waters of West Florida, and his "complaints" (as the admiral characterized them) had been reinforced by the ship captains with whom Gauld had served. Admiral William Parry had gone to considerable lengths to satisfy the surveyor's requirements, but *Charlotta* had "turned out so extremely crazey & rotten" that Gauld and his party had been "obliged to run for the Harbour to save their Lives." The "Deal cutter" was rapidly growing old in the service. Although the Admiralty apparently ordered the construction of schooners "to be built under the direction of Capt. Kennedy" in New York, and Parry hoped they would be "very near the draft of water & tunage that Mr. Gauld wants," they seem never to have been put into service on the Gulf Coast. Finally the admiral found in Jamaica a small decked-over sloop of fifty tons' burden, "drawing little more than six feet water," and answering in every respect for the surveying service. Parry ordered the naval storekeeper to purchase the vessel, and she was commissioned as the *Earl of Northampton*. Lieutenant John Pakenham raised his pendant aboard her at Port Royal on June 17, 1769, and drawing supplies and men from other vessels in the harbor, he got the little armed sloop ready for sea by July 10.

When he advised the Admiralty of these developments, Parry also reminded Their Lordships of their promise to send to Gauld "such of the instruments as were carried home by Sir John Lindsay." These

10. Lorimer to ?, June 24, 1769: Letters and Papers, Royal Society, London, V, 49–75. Gauld was in the Commons House as late as June 13 and probably remained in attendance through the sixteenth when the session was adjourned until June 23. *Jamaica* sailed from Pensacola on June 17, and Gauld was certainly aboard. Captain's log, ADM 51/980; muster tables of *Jamaica* and *Earl of Northampton*: ADM 36/7431 and 8519; captain's log, *Earl of Northampton*: ADM 51/4178. It was apparently on this voyage that Gauld met John Payne, master of *Jamaica*, with whom he struck up a close personal and professional relationship.

included a telescope, a "stop-watch" to replace or supplement one the surveyor had purchased at his own expense, "proportionable compasses" for uniformly reducing the scale of charts, and a supply of good drawing paper and lead pencils, neither of which could be obtained in Pensacola. The watch Gauld desired should be as good as the surveyor's own, which was "made by one of the best hands in London [and had] all the essential properties of Mr. Harrison's timekeeper," so that he would be able to compute observations of longitude with greater accuracy. Along with his letter Admiral Parry sent Gauld's fair copy of the previous year's work. Twelve feet long by three and one-half feet wide, this chart reflected surveys carried out in the face of great difficulties and in an open cutter, often during gales of wind. Parry believed that upon consideration of these circumstances the Admiralty would approve of George Gauld's "assiduity."[11]

Lieutenant John Pakenham in the *Earl of Northampton* moored off the town of Pensacola on August 2 but found that Gauld and *Jamaica* were gone. Replenishing his supply of victuals and water and taking on extra ballast, Pakenham left the harbor on August 8 and made his way westward, past Mobile Point and Dauphin Island, as far as the Chandeleurs. Failing to encounter the survey party, and plagued by such light breezes that he was forced to tow *Northampton* with his boats, Pakenham reversed his course, worked his way between Ship and Horn islands, and with the aid of a pilot was able, on August 13, to bring his sloop alongside *Jamaica* in Ship Island Bay. The surveyor was away from the ship when *Northampton* dropped anchor and Pakenham arranged with Captain Talbot to assume responsibility for Gauld and his party as of August 20. *Jamaica* sailed from Ship Island on the twenty-fourth, and George Gauld must have been more than a little surprised to find a new vessel at the rendezvous when he returned with the cutter on August 27.[12]

11. Parry to Admiralty, January 29, March 21, 1767, July 6, 1769: ADM 1/238.
12. Phillips to Admiralty, August 18, 1769: ADM 1/2301; captain's log, *Earl of Northampton*: ADM 51/4178. *Jamaica* was wrecked January 21, 1770, on the Colorados off the coast of Cuba. Capt. George Talbot and Master's Mate John Payne were eventually returned to England aboard *Renown*, commanded by Thomas Fitzherbert. Fitzherbert to Admiralty, July 8, 1770: ADM 1/1789; Talbot to Admiralty, February 24, July 9, 1770: ADM 1/2590. Philipps of *Tryal* protested that he was unable to visit Apalachee in August 1769 because "Mr. Gauld being absent upon the Surveying Service, I could not have recourse to his draughts." Philipps to Admiralty, October 9, 1769: ADM 1/2301. Hard experience made the captains in the Gulf of Mexico appreciate the importance of Gauld's work.

G^eo. Gauld — Cartographer

When he was finally able to return to his professional occupation, Gauld took up his work where he had left off the previous year. The cutter, provisioned for a week or ten days at a time and her crew provided with double rations of grog, enabled him to explore the inland waterways as well as the coast. The extensive notes on his field sheets indicate that he ascended the several branches of the Pearl River. He noted that all of the inhabitants had settled on the right bank of the western branch of the river and, without exception, had chosen the land on the opposite bank for their rice plantations. The first of these settlements, some fifteen miles from the mouth of the Pearl, was that of M. LaRonde; farther upstream, at intervals of a mile or two, were the plantations of LeFaber, Vincent, De la Gautrais, and Nieri. Some seven years later George Gauld himself would apply for and receive a grant on the east branch of this same river.

From the Pearl River, Gauld returned to Lake Borgne. He thought that the several creeks emptying into the lake might extend as far as the plantations on the Mississippi River, and he identified one of them as "Mason's Bayou," his phonetic spelling for Maxent's Bayou, which had been named for the wealthy French businessman Pierre Maxent, later the father-in-law of the Spanish Governor Bernardo de Gálvez. Passing on through Lake Pontchartrain, whose great size — thirteen leagues in length and two to eight leagues in width — contributed to its becoming very rough during a heavy wind and developing an unpleasant short swell, Gauld entered Lake Maurepas. On the British side of the lakes, French inhabitants engaged in cattle-raising and manufactured pitch, tar, and turpentine, which they traded in New Orleans at great profit and without interruption, for no troops or patrol vessels were stationed on the lakes to halt their illicit trade. Furthermore, Frenchmen and Spaniards, having no other source of masts or yards for their vessels, regularly took what timber they pleased from the West Florida side of the lakes. Gauld recorded only native habitation along the shores of Lake Maurepas. On the British side of the lake he found a village of Pascagoula Indians, and immediately south of the Houmas River was a village of the Biloxi tribe.[13]

While the surveyor was exploring the lakes and rivers, Lieutenant Pakenham, at Ship Island, was discovering a serious flaw in the *Earl*

13. Gauld, "Field sheets of Mississippi Sound and Lakes, 1768-1769"; Gauld, "A General Description of West Florida, 1769," pp. 5-7.

of Northampton.[14] The vessel "leaked much in her bottom," and by early September she had to be pumped every twelve hours. Twice before Gauld returned from his last outing on September 19, Pakenham heeled his ship to stop the leak, but to no avail. Shortly after the surveyor rejoined her the weather turned foul, and in the autumnal rains *Northampton's* deck leaked so badly that her people could not sleep below in their hammocks. The pumps were manned every six hours, then every four hours, and finally in desperation, on September 24, Pakenham cast off and made sail for Nassau Road. Thunderstorms continued to plague the expedition as it passed Grand Gosier and Breton islands. Then, on the thirtieth, calm followed storm, and with the flagstaff at the Balize in sight — the only way to locate the mouth of the river because of the low marshy nature of the delta — the cutter took up the task of towing *Northampton* toward the Northeast Pass into the Mississippi. Affording a depth of about ten feet, this entrance to the river was that most frequented by small vessels, and with a draught of six feet eight inches aft, *Northampton's* captain anticipated no trouble. (The East and Southeast passes were deeper, but even so, vessels drawing more than eleven and one-half feet had to lighten their loads before entering.) Pakenham crossed the bar on October 3, but in the prevailing calm it was two days before the sloop was able to stem the current, and then he was forced to warp the vessel around the river's many bends or turns — not altogether a misfortune, for he discovered that his mast was so rotten that he dare not carry full sail.

Northampton's crew were pressed to the limit ascending the Mississippi. The cutter (no doubt carrying George Gauld) went ahead sounding the river and not infrequently disappearing from sight around a bend. The ship's boats handled the warp-lines and towed when they could. Aboard, the pumps were manned hourly, and every thirty minutes by the time they reached English Turn, where the defensive batteries erected by the French were duly noted. At last, early in the afternoon of October 9, they came in sight of the town of New Orleans and the proud banner of the king of Spain. Pakenham moored his ship abreast the town. Initially he armed and

14. Captain's log, *Earl of Northampton:* ADM 51/4178, from whence the following information is derived. See also John Preston Moore, *Revolt in Louisiana: The Spanish Occupation 1766–1770* (Baton Rouge, 1976). The *Pennsylvania Gazette,* January 11, 1770, reported the survey ship's visit to New Orleans and that it had received "every assistance" and "every mark of respect . . . that could be expected."

sent two petty officers ashore to guard the moorings, but finding that the Spaniards offered no protest, he withdrew his men to the ship.

Northampton remained at New Orleans less than forty-eight hours, but in that time two seamen apparently slipped ashore; at their return they were duly punished with a dozen lashes for being drunk and disobeying orders. Rather more surprising was the receipt of a cask of pork delivered to the ship "by Order of General Oreilly." Spanish military forces led by Alejandro O'Reilly had landed at New Orleans on August 18, and while the patriotic spirit that inspired the revolt of 1768 was easily quelled, O'Reilly was still involved with the trial and conviction of the French ringleaders. New Orleans was a nervous city. The Spaniard had abruptly ejected the English merchants who had battened on New Orleans's trade for years, but he acted more diplomatically when he saw the ensign flown by the armed sloop *Northampton*. He need scarcely have troubled. With both pumps going constantly, Pakenham slipped downriver on the eleventh and barely made it past English Turn before he was forced to halt, empty the ship, and heave her over in order to caulk her bottom. After a week *Northampton* was able to proceed to the mouth of the river, and there Pakenham anchored as long again while Gauld and his surveying crew explored and charted the area. During the week Gauld was off in the cutter, a number of Spanish naval vessels came down the river, returning contingents of O'Reilly's force to Havana. On the first of November, two of the seamen off *Jamaica* took advantage of a shore party engaged in cutting wood and jumped ship. The next day Gauld returned, and *Northampton* crossed the bar and began a fast run home that put her in Pensacola Bay on November 3. Gauld's cutter—and the remaining members of his little crew—were returned to *Jamaica*.

The surveyor very probably left the ship and did not accompany *Northampton* when she sailed, November 10, carrying Lieutenant Governor Montfort Browne to Mobile. The weather was extremely nasty. Pakenham lost his jib boom and nearly lost two men with it in a squall in Mobile Bay, and when he returned at the end of the month, he was forced by great seas off the bar to wait for three days before he could enter Pensacola harbor. It took all of December to put the sloop in shape to return to Jamaica; she sailed on January 8, 1770, came into Port Royal February 6, and four days later Pakenham turned her over to Lieutenant Anthony Gibbs for complete refitting.

Service Ashore and at Sea, 1769-1771

George Gauld remained at Pensacola, discharged from *Northampton* the day she sailed.[15] For him it was time to draft charts, summarize the knowledge gained over the past five years, and become involved once more in colonial politics. His accomplishments for the year culminated in the preparation of the lengthy document entitled "A General Description of the Sea Coasts, Harbours, Lakes, Rivers &c. of the Province of West Florida." An extension and a commentary upon the surveys he had completed, the thirty-page manuscript provided a remarkably detailed and complete account of the coastal features of the colony. Drawing heavily upon his own notes and charts and only occasionally upon information derived from other sources, Gauld compiled a unique word picture of the scene in which he lived and worked. There is nothing to suggest just when he composed the "Description," and as the year was well filled, first with the General Assembly and then with his westward survey in *Jamaica* and *Northampton*, it may well have been put together in odd moments; but it reflects the unfailing precision and organization of the master cartographer.

While naval surveyor George Gauld was perfecting his charts of the coastal waterways of West Florida during the summer and fall of 1769, his military counterpart, engineer Elias Durnford, was hastening to London on a different mission. West Florida had had its fill of Lieutenant Governor Montfort Browne, already under investigation by the home government, and when Durnford met with the American Secretary of State Lord Hillsborough he undoubtedly voiced the prayers of most of the colony that Browne might be replaced. Hillsborough acted with exceptional speed. He immediately drafted orders recalling Browne and promoting Elias Durnford to his office. The new lieutenant governor (and his attractive young bride) landed in Pensacola December 26, the bearer of his own good tidings.

Word of Browne's dismissal electrified the citizens of Pensacola; they hastened to express their delight with Durnford's elevation, and George Gauld was among the number whose memorials celebrated their satisfaction with the change.[16] Durnford would have taken pleasure at reading the surveyor's signature, for he fully recognized the importance of Gauld's work for the colony and had spoken highly of him during his conferences with Hillsborough in London. A few

15. Muster table, *Earl of Northampton:* ADM 36/8519.
16. Robert E. Gray, "Elias Durnford, 1739-1794" (master's thesis, Auburn University, 1971); The Addresses of the Inhabitants of the Town of Pensacola: C.O. 5/587.

126

weeks later, when writing to the American Secretary, Durnford would refer to Gauld's charts as "very correct and explanatory." Subsequently he remarked, with noticeable professional approval:

Mr. Gauld's Plans which I mentioned to your Lordship in England are now compleat to the westward of the Mississippi and as far East as the Bounds of this Province; I must therefore beg leave again to inform My Lord that they may be had at the Admiralty office, and are really worth seeing, being surveyed with great accuracy, and neatly drawn.[17]

The new chief administrator of West Florida acted promptly to meet the colony's needs and resolve its quarrels. He summoned the General Assembly, prorogued since the end of June 1769, to take up those matters it had left unfinished during Browne's administration, and on March 1, 1770, George Gauld found himself seated once more with his fellow commoners for the second session of the Fourth General Assembly.

Gauld's initial prominence in this session of the provincial legislature may be taken to reflect his increased stature in the community as well as his personal standing with the lieutenant governor. He was chosen to wait upon Durnford at the opening of the assembly and helped to prepare the customary reply to the lieutenant governor's welcoming speech, a response in which the Commons again assured Durnford of their "peculiar satisfaction" at the opportunity of working with a man of such "abilities and goodness of heart." They disowned "the late impolitic discontinuance of taxes" as not of their choosing and promised to get on with the provincial business "without any delays or difficulties." This was, of course, the language of politics. George Gauld sought an immediate respite from his own legislative obligations and on March 3 was granted leave by the house to absent himself. His name does not reappear in the assembly minutes until March 16, but as no record of attendance was kept, he may have returned to the house before that date. The assembly kept at its work dutifully through the month of March, and on April 5, Lieutenant Governor Durnford granted their request for a three weeks' adjournment to the first of May. After the assembly resumed, Gauld was employed in working out the details of a measure to prevent debtors from fleeing the colony, and he served on a conference committee, with members of the council, that smoothed out certain

---

17. Durnford to Hillsborough, March 25, 1770: C.O. 5/587.

differences between the two houses. Along with seven other bills, including the long-delayed revenue bill, this act was approved by the lieutenant governor at the closing session of the assembly on May 19.[18]

As in the preceding year, if rather more happily, George Gauld spent the first part of 1770 in a public service other than that which he most enjoyed. When the warm, sunny days of spring returned to the Gulf Coast he undoubtedly yearned to return to his surveying activities, and recent developments at Port Royal raised his anticipation. In 1769, Admiral Parry was called home, and his place as commander in chief on the Jamaica Station was given to Gauld's former captain aboard H. M. S. *Centaur*, Arthur Forrest. From Port Royal, on August 10, 1769, Commodore Forrest wrote to inform Gauld of his arrival in Jamaica and to express the hope that "you go on prosperously in your surveys." While bringing to Gauld's attention the desirability of acquiring a thorough knowledge of "the Dry Tortugas, Martyrs and round Cape Florida . . . If a good Harbour could be found there, it would be of great service," Forrest sagely left Gauld to determine his future course for himself. "Until I know what you have done, I cannot pretend to give you any directions," and with a touch of both personal and professional solicitude he concluded, "you'll therefore write me by every opportunity."[19]

The surveyor completed the drafting of a fair copy of his recent investigations early in the winter and dispatched it to Forrest about January 7, 1770. In his letter of transmittal Gauld raised an intriguing point relative to the boundaries of West Florida. By the Treaty of Paris, he argued, France had ceded to Britain all lands east of the Mississippi River save the Isle of Orleans. The northern boundary of that very ill-defined area was demarked by "the River Ibberville (which is dry for several Miles a great part of the year)." Sensibly assuming that an "island" should be bounded by the nearest waters surrounding it, Gauld reasoned, "certainly that large Branch of the River at the Pass a La Loutre (where there is constantly deep water and a rapid stream) ought with propriety to be looked upon as the other Boundary." By Gauld's reasoning, "the new Balise, erected by the Spaniards, is without the Boundary of the Island of New Orleans, to which alone they are limited by the Treaty of Peace." Such an interpretation would give Britain claim to that portion of the Mis-

18. Rea and Howard, *General Assembly*, pp. 206–43.
19. *An Account of The Surveys of Florida*, p. 5.

sissippi delta containing the modern North and Northeast passes — in effect it would divide the entrances of the river between Britain and Spain, to the great and obvious advantage of Britain. Gauld carefully noted these waterways on his chart, but the surveyor's technique for encroaching upon Spanish Louisiana was not pursued by Arthur Forrest. The commodore acknowledged Gauld's scheme but merely replied that the Admiralty should be advised of it "at a proper time."[20]

The commander in chief at Port Royal was concerned with other problems. The West Floridians, more excited by dangers from hostile Indians than he thought they need be, were pressing for increased naval support, even though Forrest had already stationed the sloops *Stag* and *Druid* at the provincial capital. The *Earl of Northampton* had returned to Jamaica "in want of everything" and would require two months of refitting. As she was on special assignment as Gauld's survey vessel, it was necessary to equip, man, and provision her at the expense of the other ships in the harbor. Forrest proposed that she be placed upon a regular establishment, thereby relieving the rest of the squadron of supplying her and, at the same time, providing the coastal survey with a more efficient ship in which to operate. The subsequent extended service of *Northampton*'s complement suggests that the Admiralty approved of Forrest's plan, and as a result Gauld was able to begin each subsequent season with at least part if not all of an experienced crew.[21]

Although George Gauld failed to change the political map of North America by dint of cartographic technicalities, his accomplishments did not pass unnoticed elsewhere in the colonies. Recognition and honor of the highest order came to him in 1770. At the meeting of the American Philosophical Society held in Philadelphia on January 19, the "surveyor of W[est] Florida" was elected to membership in that most distinguished body of scientists, of which Benjamin Franklin was president. Among the other nine gentlemen brought into the society that day were Dr. John Fothergill, Edward Nairne, James Ferguson of London, a professor of chemistry at the University of Utrecht, and that remarkable character Lord Stirling of New Jersey. The bylaws of the American Philosophical Society re-

20. Gauld to Forrest, January 7, 1770; Forrest to Gauld, March 29, 1770, extracts in Chester to Hillsborough, June 7, 1774; C.O. 5/591, fol. 257.

21. Captain's log, *Earl of Northampton*: ADM 51/4178; Muster table, *Earl of Northampton*: ADM 36/8519; Forrest to Admiralty, February 28, 1770: ADM 1/238.

quired that the name of a candidate for membership be posted "for the view of the Society" prior to his election. Therefore Gauld's name would have been presented to the members at their meeting of December 15, 1769; the secretaries were directed to notify him of his election at the next meeting following his acceptance in January.

George Gauld was not the first West Floridian to gain membership in the American Philosophical Society. That honor was shared by the surveyor's good friend Dr. John Lorimer and, for less obvious scientific reasons, by Daniel Clark, an early settler and sometime member of both the Commons House and governor's council. They had been elected to the society April 21, 1769, and it was undoubtedly due to John Lorimer's affection and respect for Gauld that the surveyor's name was presented to the society. Lorimer, who had an interest in the astronomical aspects of establishing longitude and would publish a book on terrestrial magnetism, was familiar with Gauld's work and repeatedly brought it to the attention of his correspondents in the more northerly colonies. The army doctor also acted as a sort of unofficial clearinghouse for several cartographers who worked in West Florida in these years and kept Gauld up to date on the contributions of Thomas Hutchins and Bernard Romans, both of whom would also find a place in the list of American Philosophical Society members.

Gauld's association with the society was more honorific than productive. In 1773 he sent to Philadelphia, through Dr. Hugh Williamson, a few items of random interest: "A draught of Chester & Middle rivers in W[est] Florida" made by Thomas Hutchins, his own "Measurement of the height of Blue Mountain & Catherine Hill" in Jamaica, and the more significant "Description of the Coast of West Florida." With these he included, as curiosities of natural history, a shark's jaw, the skin of a shagreen fish, and a porcupine fish. At the same time Gauld made a cash contribution of two half-Johanneses to the society, apparently the only dues he ever paid.[22]

The "Description of the Coast of West Florida," which Gauld undoubtedly hoped to see published in the society's *Transactions*, was delivered to the secretary, Owen Biddle, by Dr. Williamson on August 20, 1773. It did not win the favor of its editorial reader and was rejected with a laconic comment penned on the covering sheet: "This

22. *Early Proceedings of the American Philosophical Society* (Philadelphia, 1884), pp. 46–48, 80–82, 168, 343–44; *Transactions of the American Philosophical Society* (Philadelphia, 1771), pp. vii, xiii; Rea and Holmes, "Dr. John Lorimer," pp. 367–69. The Johannes was a Portuguese gold coin that circulated in the American colonies.

> *A General Description of the Sea Coasts, Harbours, Lakes, Rivers &c.ᵃ of the Province of*
>
> *West Florida 1769*
>
> *by George Gauld*
>
> The Province of West Florida the Frontier of the British Dominions in America, includes a space of about 120 Leagues, on the North side of the Gulph of Mexico, from the River Apalachicola (the Boundary between it and East Florida) to the Mouth of the Mississippi; the Island on which the Town of New Orleans stands only excepted.
>
> To enter into a particular Detail of the Advantages that may arise to Great Britain in time, from the Situation and Productions of this Province, may perhaps be foreign to the present purpose; though as far as they are connected with the Interest of a Trading and Commercial Nation, and particularly a Maritime Power, I shall just occasionally mention such of them as to me seem most deserving of Notice.
>
> *General Observations*
>
> With regard to the Coast there are some general Observations, that may be of Service in making the Land; which is distinguishable several different ways; as by the Latitude, the Trenchings and Direction of the Shore, and the Soundings, and Quality of the Bottom.
>
> From Cape Blaize in 29°.41′ N° Latitude, to the Balize at the Mouth of the Mississippi, the Coast forms a Curve, inclining to the Northward for 28 Leagues, as far as the East end of Rose Island in 30°.28′ N°. from thence the Land gradually declines to the Southward, as far as Mobile Point in 00°.17′ N°; about 13 Leagues; Dauphin Island and the rest of them including Ship Island, stretch nearly West for the space of 20 Leagues, and from the N° end of the Chandeleurs, which lyes near 5 Leagues to the SE of Ship Island, the Coast runs chiefly to the Southward to the Entrance of the River Mississippi.
>
> It is likewise to be observed that in several places, there is double Land to be seen over the different Bays, and Lagoons; as at S⁺ Andrews Bay, which may be known by a high white Sand Hill near the Point of a Peninsula on the Left hand going in; at S⁺ Roses

IX. "A General Description . . . of West Florida." This is the title page of Gauld's meticulous description of "the Sea Coasts, Harbours, Lakes, Rivers &c.," which provides a summary of his observations after working for five years along the Gulf Coast. Its contents and its phraseology were freely borrowed by Bernard Romans and Thomas Hutchins. (Courtesy of the American Philosophical Society.)

long uninteresting Paper can hardly obtain a Place in the Transactions of a Philosophical Society. It should however be preserved in the Files for the use of Historians or Map Makers." The judgment was not altogether unfair or, for that matter, derogatory, for George Gauld was no philosopher in the modern or scholarly sense of the word; rather he was a highly skilled professional in a field that had yet to gain much scientific standing—though that day was not far distant. Nevertheless, Gauld's "Description" was put to use by the society just a month after it was submitted. Bernard Romans exhibited his "Chart of the Navigation to, & in, the Ceeded Countries" to the American Philosophical Society, and a committee of five members seized upon Gauld's "Description" as a basis for comparison and evaluation. In light of Romans's later productions and his debt, acknowledged or otherwise, to both George Gauld and John Lorimer, there is a certain irony in the fate of Gauld's manuscript which was and is duly "preserved in the Files" of the society.[23]

The minutes of the American Philosophical Society meeting on January 21, 1774, indicate that George Gauld was elected to membership for a second time on that occasion and that among the other inductees was Bernard Romans. Since membership was conferred for life, this action by the society presents a minor mystery. It may be resolved by noting that the "Laws and Regulations" of the society required the payment of *Ten Shillings* admission money," and George Gauld may well have been a delinquent whose reinstatement was treated as a new membership. His was not the only such case. Although the society required a similar annual payment from members who enjoyed voting privileges, only residents of Philadelphia were expected to attend and excercise that right, and few of them seem to have done so. The explanation of Gauld's second election may simply lie in the division of duties and responsibilities among four society secretaries.[24] Learned and scholarly societies are not always blessed with efficient officers, even in the twentieth century.

With the completion of refitting and the advent of fair weather in the Gulf of Mexico, Lieutenant Anthony Gibbs prepared to put the

23. Gauld, "A General Description of West Florida, 1769," is listed in Whitfield J. Bell, Jr., and Murphy D. Smith, *Archives and Manuscript Collections of the American Philosophical Society* (Philadelphia, 1966), pp. 58–59, and is preserved in the society's archives as MS 917.59: G23.

24. *Early Proceedings*, pp. 74, 86–87; *Transactions, 1771*, pp. v–vi; P. Lee Phillips, *Notes on the Life and Works of Bernard Romans*, ed. John D. Ware, facsimile ed. (Gainesville, 1975), p. 49; letters from Whitfield J. Bell to John Ware, July 29, October 18, 1971.

Earl of Northampton into service again.²⁵ On April 14, 1770, the surveying sloop stood out of Port Royal harbor. Gibbs made his landfall off Mobile Point and anchored at Pensacola on May 6. The assembly was still in session, so it was a fortnight before Gauld could turn his attention to the summer's work. At last, on the thirtieth of the month, the party was ready and *Northampton* sailed, Gauld's twelve-oared cutter in tow. Off Santa Rosa Island, the next day, they encountered a sloop from Guinea transporting African slaves to Pensacola, a promising sign of West Florida's improving economy. On June 2, *Northampton* entered and moored in St. Joseph Bay. The cutter was promptly put to work as Gauld renewed his study of that body of water, but between the thirteenth and eighteenth Lieutenant Gibbs found it necessary to give the cutter a thorough overhauling.

On June 20, the little expedition set out from St. Joseph Bay, the surveying cutter in tow again, and the following day Gibbs cast her off to sail ahead, for "she sail'd much better than the Sloop." *Northampton* proceeded slowly eastward, sounding regularly, and during a period of at least forty-eight hours her watch saw no sign of the swifter vessel, which had disappeared over the horizon. By the twenty-fifth Gibbs reached the western coast of East Florida and began working southward, lying in or well off the shore as wind and water dictated. The ship's normally quiet routine was rudely interrupted on the twenty-seventh, and the potential dangers of coastal observation were firmly impressed upon all hands. Moving in toward the shore south of Cedar Keys, Gibbs suddenly, without warning, found himself on the edge of Seahorse Reef. Veering off at the last moment, he dispatched Gauld in the ship's boat to examine the shoal, and at the same time he angrily ordered a dozen lashes for seaman William Jackson "for neglect of duty, for not seeing the shoal as he was at the mast head to look out for those things." Henceforth Lieutenant Gibbs would be extremely wary of the Florida coast. The next day the sloop anchored in what Gauld called "the Bay of Amasura," now known as Waccasassa Bay, and the surveyor, whose boat had followed the coastline more closely than had the ship, returned for supplies. As Gauld was unable "to find a passage into the Amasura . . . owing to the shallowness of the water," they proceeded southward with the boat scouting ahead and a sharper watch at the masthead.

25. Captain's log, *Earl of Northampton:* ADM 51/4178.

Service Ashore and at Sea, 1769–1771

On July 2, *Northampton* sighted Egmont Island, and approaching the entrance cautiously, with an eye on Gauld's earlier soundings, Lieutenant Gibbs sailed into Tampa Bay on the third, by way of the Southwest Channel. For the next ten days *Northampton* remained in Espíritu Santo, moving occasionally, sending crewmen ashore for wood and water, while the surveyor and four men were out in the ship's small boat rechecking and adding to Gauld's cartographic details. Lieutenant Gibbs observed that the water on Egmont Island, which he "obtain'd by sinking well," was "very good." He would have cause to think often of it during the weeks to come.

Northampton stood out of Tampa Bay on the fourteenth of July and began running southward, roughly paralleling the shore. Understandably nervous after nearly going aground in June, Gibbs lay well out to sea, and daily squalls during the last week of July kept him from engaging in any close observations. By the end of the month he had passed Charlotte Harbor, Sanibel Island, and Cape Romano but without noting anything of interest. The weather, which continued to be squally, kept Gibbs at a distance from the shore, which must have caused George Gauld considerable frustration. For his part, Gibbs was growing concerned over the ship's diminishing supply of drinking water; on the tenth of August he began to ration the crew to three quarts per day. By the middle of the month *Northampton* was off Cape Sable, and from "Point Janchia" (Gauld's Punta Tancha) Gibbs began sounding in toward the shore, looking for an opportunity to refill his casks. The water was very shoal, forcing him to anchor at nightfall, and sudden squalls frequently threatened to drive him aground. On the nineteenth he spread a sail to catch rainwater, but the skies could not possibly supply the ship's needs. Then, on the twenty-first, they sighted the Florida keys stretching in a line from southwest to east northeast; Gibbs's hopes rose, but he prudently cut the sloop's water ration down to two quarts a day. There followed a thirsty, unrequited search, island by island, until the twenty-seventh when "the Boat return'd having found some water too brackish to use." Running back to the mainland and himself going ashore to lead the hunt for water, it was not until September 1 that Lieutenant Gibbs "found some old wells, the water better . . . but brackish," and anchored for ten days to complete watering. Gauld used the opportunity to get out with the boat "to take some bearings." Regrettably, the results of his plotting were not recorded, and Gibbs had only a general idea of his ship's location. From the tenth to the twentieth of September, *North-*

G^(eo.) Gauld — Cartographer

ampton worked her way westward along the keys, anchoring abreast one, "a high white Island, to try for water, what we have being very bad." The log recorded the discovery of "many Islands, as we stand to the Westward," but no water. On the twenty-first of September, Gibbs reported that he lay off "Key Marquet," the Marquesas, but he was still without water, and on the twenty-seventh *Northampton* endured an exceptionally hard blow in which a cable parted and she lost an anchor. Having only one cable left, and that patched together out of two hawsers, Gibbs and Gauld consulted on the problem they now faced — that of working in shallow water in which they must be able to anchor the ship securely if they were to avoid being grounded by an onshore gale. "It is the Surveyor's Opinion as well as mine," Gibbs recorded in his log, "nothing Farther can be attempted for the Service this Season, therefore it will be best to return to Pensacola to refit the Sloop." On the twenty-eighth Gibbs swung north of the Dry Tortugas, where he spoke the *King George* bound for Philadelphia from the Bay of Honduras, and set a northwesterly course for home. On October 6 he identified the land ahead as Santa Rosa Island and bore away for Pensacola. George Gauld completed his voyage and went ashore on the eighth. Gibbs took *Northampton* across to the careening wharf at Deer Point, and there he discovered that the ship's bottom was "greatly eat by the Worms, & part of the false keel gone," not to mention many rotten timbers fore and aft. It was Christmas Day before the sloop was in sufficient repair to undertake the voyage back to Port Royal, and she would require nearly four months of work in the docks before she was ready for duty again.[26]

The winter of 1770–71 passed quietly for George Gauld. A new governor, Peter Chester, had landed at Pensacola while the surveyor was cruising in *Northampton* at the far distant end of East Florida. Governor Chester was not a man to act hurriedly, nor did he feel any great pressure to summon a new General Assembly. Not until April 1771 did he decide to hold elections, and then he announced that the assembly would not meet until June. These circumstances undoubt-

26. After returning to Port Royal in 1771, Lieutenant Anthony Gibbs was assigned to *Sir Edward Hawke* and sent on a mysterious mission to Rosario Island, where he was intercepted by two Spanish *guarda costas*, taken to Carthagena, interrogated, and then released. Admiral George Rodney extracted an apology from the Spanish governor at gunpoint; he also ordered Gibbs court-martialed for surrendering his ship to the Spaniards without firing a single shot. Poor Gibbs was found guilty and dismissed from the service. See David Spinney, *Rodney* (Annapolis, 1969), pp. 252–61.

Service Ashore and at Sea, 1769–1771

edly led Gauld to forego the resumption of his career in politics and to devote the year to surveying.[27] In his professional endeavors he continued to enjoy the highest respect. Lieutenant Governor Durnford, who also returned to his trade after Chester took the reins of the colony, remarked most favorably upon Gauld's achievements. Not only was Gauld exceptionally competent; he was generous in sharing his knowledge and his charts with other surveyors, whatever their interest might be. In fact, Durnford admitted, in all of West Florida, "no actual Surveys except along the Sea Shore have been made and those by Mr. George Gauld Naval Surveyor only."[28] Governor Peter Chester would discover those talents for himself, before long, but at present he left Gauld to his own devices.

The surveying sloop *Northampton*, refitting in Port Royal, received a new commanding officer in the spring of 1771. Acting Lieutenant Walter Anderson went aboard to see the ship's repairs and provisioning completed, and on May 21 he got her to sea, the surveyor's boat in tow, and moored in Pensacola harbor June 13.[29] There was still work to be done — stores for a four-month voyage to be loaded and the small boat to be caulked. On June 20, George Gauld went aboard, and *Northampton* stood down the bay.

Gauld was clearly a cartographic perfectionist whose curiosity and professional demands could only be satisfied by the most intensive coverage of any piece of ocean bottom or shoreline. Sailing from Pensacola harbor, *Northampton* proceeded out to sea, rather than striking eastward, in order to extend Gauld's soundings on the bank outside the bar. That operation was interrupted by a squall that cost the ship her topmast, and no doubt on Gauld's advice Lieutenant Anderson put into St. Joseph Bay on the twenty-third to cut a new mast. Gauld probably took advantage of the opportunity to add to his store of descriptive details during the week the sloop remained at anchor. Finally, at sunrise July 1, *Northampton* swung out of the bay and made for Anclote Island, raising her landfall on the third, then turning down to enter Tampa Bay on July 5.

27. On Chester see Eron D. Rowland, "Peter Chester," *Publications of the Mississippi Historical Society*, Centenary Series, 5 (Jackson, Miss., 1925); Lucille Griffith, "Peter Chester and the End of the British Empire in West Florida," *Alabama Review* 30 (1977); Rea and Howard, *General Assembly*, pp. 244–45.

28. Elias Durnford to John Ellis, June 24, 1771: Treasury Papers T 1/475 (P.R.O.); Durnford to Hillsborough, August 31: C.O. 5/588.

29. The description of the 1771 survey is based upon the captain's log, *Earl of Northampton*: ADM 51/4178.

G^{eo.} Gauld — Cartographer

George Gauld could scarce get his fill of the great Bay of Espiritu Santo. He went ashore immediately after *Northampton* anchored and soon began working northward up the bay. About the end of July he seems to have made an excursion along the Gulf Coast to Anclote and other islands outside Tampa Bay, but by August 11 he was back and working in Hillsborough Bay, where Lieutenant Anderson anchored the sloop for the surveyor's greater convenience. Aboard ship they were "employed as usual," thought it excessively hot, and they found the water ashore unpalatable until they began to draw their supply from the Hillsborough and Palm rivers at the head of the bay. By the twenty-second the surveyor's boat was in need of caulking and minor repairs, but these were completed in forty-eight hours; and on the twenty-sixth George Gauld and nine men went surveying southward, carrying three weeks' provisions and the double ration of rum that was customarily allowed the boat party.

Gauld was gone for exactly one month. No doubt, like the crew remaining behind on *Northampton*, he and his men stretched their rations by pretty regular fishing. For him, however, it was an absolute necessity, for upon his return, September 26, he reported that the boat had struck upon a submerged tree stump and foundered in five feet of water. All of his provisions and stores had been lost. Shipwreck was a hazard to be taken in stride, however, and by October 7 the surveyor and a crew of nine were off to the "West River" (perhaps the Alafia) for a week-long investigation of that side of the bay. Gauld returned on the fourteenth, and three days later his boat followed *Northampton* down the bay and out the South Channel.

Finding the seas off Egmont and Anclote islands too rough for working the ship's boats, Gauld and Anderson dropped south of Tampa Bay to "San Carlos Bay" — Charlotte Harbor — where they anchored on October 21. There *Northampton* would remain until November 29. Gauld's survey got under way by October 23, when Lieutenant Anderson sent some of his people ashore to plant three flags which the surveyor would use as markers, while Gauld and his customary crew of nine began working up the bay, carrying a fortnight's provisions — and double rum. Unfortunately the weather remained foul. The boat returned two days later, "much Damaged by the Bad Weather." She was hauled ashore for repairs, which were not finished until the thirtieth. The work crew returned to the ship each night. On the morning of the twenty-seventh, while going into a boat to go ashore, carpenter Robert Spence fell overboard and drowned.

Service Ashore and at Sea, 1769-1771

Lieutenant Anderson had other problems as well. The ship's provisions were running low, and he had put the crew on half-rations of bread, beef, and pork as soon as he anchored in Charlotte Harbor. By early November he was regularly sending a party oystering, and he noted in his log for November 15, "our Butter all Expended." The surveyor's boat was patched up in less than a week, but the great surf on the beach made it impossible to launch her for several days. Then she proved still very leaky and was not finally ready for Gauld's use until November 3. He promptly took her out with a reinforced fourteen-man party to survey the entrance and the shoals off the harbor, more than a week's task. It was not, in fact, until the twenty-third that Lieutenant Anderson had an opportunity to use the boat to investigate some smoke to the south which he had first seen a few days earlier. The fires proved to be those of the crews of three Spanish schooners that lay at anchor in a nearby bay. The interlopers came from Havana and were engaged, as was fairly common, in fishing on the East Florida coast. Gauld accompanied the British reconnaissance party, "went ashore to the Spanish fishing stages, found there 9 people, and 3 or 4 snug palmetto huts, and plenty of carp and other fish on hooks, a dressing on the stage." He noted that:

They begin by pressing the fish with a great weight after it is split and salted, then hang it up to dry . . . : the last operation is, to pile it up in the huts ready for loading. They supply the Havanna, and the other Spanish settlements in the West Indies, in the Lent season, in the same manner as New Foundland supplies those in the Mediterranean. It is a very lucrative branch of trade.[30]

Anderson was clearly anxious to be off. He brought in the surveyor's flags on the twenty-fourth and hauled the survey boat alongside on the twenty-seventh, preparatory to sailing, only to be becalmed until the twenty-ninth. He then headed directly for Pensacola, calculating nautical miles against a rapidly dwindling food supply. On December 8 he distributed the last of the rum and on the tenth the last of the beef, but that day he also sighted the entrance to Pensacola Bay off his port bow and sailed straight into the harbor without troubling to pick up a pilot. The exactness of his navigation was a tribute to both his own seamanship and to George Gauld's car-

30. Gauld, *Observations on the Florida Kays, Reef and Gulf* (London, 1796), pp. 25-26n.

tography; together they were rapidly turning the eastern Gulf of Mexico into a neatly charted British lake.

Gauld went ashore at Pensacola to discover that, in his absence, orders had arrived that would affect all hands aboard the surveying sloop. Lieutenant Anderson was superseded in command of the *Earl of Northampton* by Lieutenant James Francis Edward Drummond, and it was he who saw her careened and cleaned at Deer Point and, on December 24, took her back to sea. George Gauld accompanied his ship to Jamaica, for there were also orders for him from a new commander in chief at Port Royal who had need of a naval surveyor's skill in his own harbor. Gauld's orders brooked no delay and went straight to the point — as was characteristic of Admiral Sir George Brydges Rodney; he would exchange the quiet of the winter in Pensacola for the bustle of Kingston and Port Royal.

CHAPTER IX

Jamaica to Key West, 1772-1773

On January 26, 1772, George Gauld returned to Jamaica, the nerve center of naval activity in the Gulf of Mexico and the Caribbean. The harbor that was home to a British squadron had changed little since Gauld first saw it in 1764. The navy's base, Port Royal, lay at the foreshortened end of a nine-mile-long sandspit, known as the Palisadoes, that closed the harbor on the south and protected the town of Kingston.[1] The broad mouth of the bay was guarded here by Fort Charles and on the northern shore by Fort Augusta. Beyond Fort Augusta and Mosquito Point lay Hunt's Bay, and back behind the "sandy hillocks" connecting Port Royal to the mainland lay the expanse of Kingston Harbor. Torn from the grasp of the faltering Spanish empire in the mid-seventeenth century, Jamaica — most particularly Port Royal — was Britain's answer to Bourbon domination of the Antilles from Cuba, Hispaniola, and Puerto Rico. Small, isolated, encompassed by enemies it was; but its planters made fortunes from sugar by the sweat of the slave population, and the Royal Navy provided its security, for the troops on the island were scarcely sufficient to overawe the blacks, not to mention the scarcely subdued maroons.

The Jamaica squadron provided protection for British shipping in the Gulf of Mexico, and in time of war, convoys of merchantmen returning to England were formed at Port Royal. Here George Johnstone and John Eliot had won the naval honors they doffed to become governors of West Florida. Nearly every ship bound from England to Pensacola stopped at Port Royal, as did *Tartar* in 1764, to replenish water and supplies. A regular packet service linked Jamaica and West Florida, though not as frequently as Pensacola merchants deemed desirable. To Kingston they shipped their lumber, and from Jamaica they took the rum that would lubricate the trade in skins and furs in Mobile, Manchac, and Natchez.

Command of the Jamaica Station was an honor but also a burden.

---

1. See Michael Pawson and David Buisseret, *Port Royal, Jamaica* (Oxford, 1975). The northern part of Old Port Royal slipped beneath the sea during the devastating earthquake of 1692.

G^{eo.} Gauld — Cartographer

Port Royal's responsibilities were greater than its resources; it could refit a ship but not rebuild one. Just as captains at Pensacola limped back to Jamaica to recover their little vessels from damages wrought by wind, sea, and worm, so the admiral on station sent his great ships in relay back to Portsmouth, praying that they would survive the Atlantic crossing — and some did not. Strong ships fell prey to the elements; strong men sickened and died of fevers that enfeebled, if they did not kill, even the best, even a Nelson. Sir William Burnaby had feared for his life before he was recalled from Port Royal; Arthur Forrest lost his within a year after assuming command at Jamaica. And it was all so deceptively beautiful as one's eye passed over the blue waters of the bay, across the white buildings of Kingston, up the lush green valleys to the hazy Blue Mountains beyond — a paradise flawed by nature, not yet improved by man.

To Admiral Sir George Brydges Rodney (who would protect his own health by retiring to the mountains during the worst of the heat) the immediate problem was that after a hundred years and more, the Royal Navy still lacked accurate charts and markings of the bay, and His Majesty's ships frequently came to grief in their own great harbor. The sixty-gun *Achilles* had gone aground on Middle Knoll, in the center of the Eastern Channel, only a half mile off Port Royal Point, as recently as December 1771. Sir George would correct that situation, as he would others at Port Royal; he would buoy the shoals and make the channels safe, and George Gauld would show the way by surveying the best-known waters in the British West Indies. Rodney explained it to the Admiralty:

I shall employ him in surveying this Harbour and the shoals at the mouth of it, as likewise the Coast of this Island till the season will permit him to resume the Survey of the Florida Coast. It is absolutely necessary that a new survey should be taken of the Channels leading into this Harbour as there is scarce a Pilot who is capable of taking charge of a Ship of War either coming in or going out, owing to the shifting of the Sands.

As a pointed reminder of the seriousness of the problem, Rodney advised his superiors that he had just court-martialed and broken the pilot who had run H. M. S. *Achilles* aground.[2]

---

2. Rodney to Admiralty, January 29, 1772: ADM 1/239. Also see David Spinney, *Rodney* (Annapolis, 1969), pp. 252–61. For an example of the poor charts available at this time see Capt. Joseph Speer's "A Plan of the Harbour of Port Royal" in *The West Indian Pilot* (London, 1766).

Jamaica to Key West, 1772-1773

George Gauld had his work cut out, work he must have enjoyed after the trials and frustrations of the Florida coast. At Port Royal he had access to all the supplies and all the equipment he might need. Engaged in the admiral's business, he also enjoyed the cooperation of every naval ship and facility. He had a harsh taskmaster, but he would fulfill his requirements, provide the Admiralty with a magnificent chart of the harbor, and fall short of the goal of adequately safeguarding the waters of Kingston and Port Royal only because Rodney himself failed to carry the project to completion.

Evidence does not indicate just where or how Gauld began his task at Port Royal. The *Earl of Northampton* required considerable refitting in February and March 1772; perhaps at this time the surveyor laid his plans and schedule, familiarized himself with existing charts, and made his own observations from the land. His ship does not appear to have been involved in surveying activity until April 1, when a new captain, Lieutenant Charles Knatchbull, who went aboard March 11, borrowed a yawl from another naval vessel in the harbor for Gauld's use.[3]

The first week of April, Knatchbull took his ship out to East Key, four miles off Port Royal Point, and moored there while Gauld worked out of H. M. S. *Boyne*'s yawl. The latter part of the month, after a brief return to the harbor, *Northampton* was moored so that Gauld could conveniently sound and chart the area between South and South East keys. From these outlying islets the surveyor worked west across the approach to the harbor, while his ship anchored off One Bush Reef and then in Hellshire Bay. Other ships' boats were pressed into service from time to time, and it is evident that Gauld was directing a considerable party in his work, for men were drafted to assist the surveyor from nearly every naval vessel in the harbor. Among these men was John Payne, formerly of *Jamaica*, who joined *Northampton*'s muster as a supernumerary from *Princess Amelia* on June 20 and thus began a long connection with George Gauld and the southern coasts of British America. During the latter part of May, Gauld moved south and west, farther out in the approaches, to sound Wreck Reef and chart a safe course into the harbor from the southwest. The eastern approach was charted, between East Key and Plumb Point, in June and as far as Rackham's Key by early July.

3. The order of Gauld's work is suggested by *Northampton*'s log: ADM 51/4178.

After July 19, Gauld worked inside the harbor area, first north into Hunt's Bay, then, a month later, in Kingston Harbor itself. Active surveying was apparently completed in the fall, for by December Gauld had consolidated his findings into finished charts. *Northampton* was designated as transport ship for a company of the 60th Regiment, which relieved a detachment of the 66th at Port Morant early in December, but it is unlikely that Gauld sailed with her on that brief run. More likely he accompanied Lieutenant Knatchbull when he sailed, December 20, in company with H. M. S. *Merlin*, east around Morant Point, then northeast for the Windward Passage. After an overnight stop at Mole St. Nicholas, the two vessels proceeded past Great Inagua and into the Caicos Passage before they parted and *Northampton* reversed her course to return, on January 14, to Port Royal. A week later Knatchbull was replaced as commander of *Northampton* by Lieutenant Nathaniel Phillips and sailed for home.

George Gauld's Jamaican interlude lasted longer than Admiral Rodney had believed (or would have had the Admiralty believe) would be necessary. One must wonder if both Georges did not recognize that the meticulous charting of the waters of Port Royal would take more than a few months. Gauld had scarcely begun his work in March when Rodney advised the Admiralty that he was busy surveying "this Harbour and the many Shoals at its Entrance." "Mr. Gauld will have compleated the Survey before the Time necessary for his return to Florida," Rodney confidently predicted. Whether he implied a return in the summer of 1772 or later, it was the following year before Rodney released Gauld to resume his official duties. In any case, the admiral seems to have been well pleased with Gauld's work and supported it at home by projecting great things for the future.

The special schooners designed for survey work had finally appeared at Port Royal, and in March 1772, Rodney assured the Admiralty that "at present they are employed in attending Mr. Gauld on his survey." (If so, there is no evidence of it in *Northampton's* log!) When Gauld had finished his task at Port Royal, Rodney continued, "I propose to send the Zephyr, the Egmont Schooner and the Northampton Sloop to assist him in the Survey of the Coasts about the Florida Cape." After that, if Their Lordships approved, the admiral suggested sending Gauld with the sloop and a schooner "to resurvey the Mouths of the Mississippi, and afterwards to proceed up the said River as far as they can possibly go."[4] That project sounds rather more like the dream of George Gauld than the long-range plan of Admiral Rodney; Gauld would indeed make a Mississippi River trip, but not at the behest of the admiral.

"For Sir George Brydges Rodney, Bart., Vice Adml. of the Red, & Comm. in Chief at Jamaica &c. &c." Gauld prepared, during the last months of 1772, a great chart entitled "A General Plan of the Harbours of Port Royal and Kingston, Jamaica, with the Channels leading thereto, and the Kays & Shoals adjacent; including Wreck Reef &c." A second version, "A General Plan of the Harbours of Port Royal and Kingston, Jamaica, with the Channels leading thereto, and the dangerous Shoals adjacent, including Wreck Reef &c.," was prepared "For the Right Honble. The Board of Admiralty" and bore the date December 1772.[5] This last chart was supplemented by extensive "Remarks and Directions concerning the Channels into Port Royal and Kingston Harbours, with Cautions to avoid the Dangers that ly in the way," which George Gauld obviously thought would be of great value to British mariners and which he deemed worthy of publication.

The "Remarks and Directions" begin with the observation that "Nature has formed a Bulwark of Reefs, Shoals and Kays, or small Islands, without the entrance of Port Royal Harbour, which makes it of difficult access to strangers, and at the same time is a great defence and Security against the Sea; thereby rendering it one of the best and safest Harbours in the W. Indies." Gauld then described the two principal channels, the eastern and southern, the navigational marks to be observed in using them, and the dangers to be avoided. While instructing his reader how to avoid the first shoal outside the harbor entrance, Gauld remarked that as Admiral Rodney "has ordered Buoys to be placed on this and all other Shoals where there is the greatest danger, the Navigation in and out of the harbour will be ren-

---

4. Rodney to Admiralty, March 22, 1772: ADM 1/239.
5. MODHD A640 Ag 4 and D961 16n. The exact date of the month is obscured by a tear in the manuscript. A manuscript copy of the "Directions for the Harbours of Port Royal & Kingston" signed by Gauld December 20, 1772, is in the Rodney Papers, P.R.O. 30/20:18. Gauld's work is briefly noticed by Pawson and Buisseret, *Port Royal,* pp. 134–35, and they reproduce a panoramic view showing "Leading marks for the south Channel" (Pl. 18) which "may well be in Gauld's hand."

dered much more easy and convenient."[6] In fact, Middle Knoll in the eastern channel, where *Achilles* grounded in December 1771, had already been buoyed when Gauld composed his "Remarks."

"In coming to anchor in Port Royal Harbour," Gauld wrote, "you must take care to give the Hospital a berth of near a Cable's length, to avoid the Shoal where Fort James, and a great part of Old Port Royal, was sunk in the memorable Earthquake 1692." The surveyor noted that there was five feet of water on the shoal in 1772 and that "it is mostly covered with Coral, which being of a vegetative nature, seems to encrease fast. . . . The Coral is a very curious natural production, being a kind of vegetable Stone, growing under water, with branches resembling Stag's horns, some broad and spreading, others small and pointed, and regularly disposed on the principal Trunk."

"The Harbour of Port Royal is spacious, with very good anchorage where no common winds can hurt, being well secured from the Sea by the Peninsula on which Port Royal stands," Gauld observed, but he went to great lengths to indicate the many hazards in the harbor and the marks by which they might be circumvented. As recently as June 1772, H.M.S. *Carysfort* had gone aground on Pelican Spit while attempting to negotiate the narrow deep-water channel leading up the bay to Kingston, and that channel had been buoyed in the course of Gauld's work. Still more buoys marked the safe channel on to Kingston where there was "deep water close to the Wharfs . . . and good anchoring ground . . . a very extensive and excellent Harbour."

Having described the route into the harbor and up to Kingston by way of the East Channel, Gauld then traced the course out of the bay via the more difficult South Channel. In describing it Gauld noted the locations of two more buoys which promised to make this approach far safer than in the past and more convenient for shipping coming in from the west or departing eastward.

Both of Gauld's charts included detailed plans of Port Royal's town and "Fort Augusta and the Narrow Channel going up to Kingston." The surveyor's "Remarks and Directions" for the Admiralty were supplemented by a fine set of panoramic views of the area—Port Royal and its environs as a pilot aboard an approaching vessel would

---

6. Sections of the old mainmast of the flagship *Princess Amelia* were used as buoys and were fastened in position with old chains and anchors. Spinney, *Rodney*, p. 261.

see it. The most striking illustration is the "View of Port Royal taken from Rackam's Kay," a visual representation of the town plan drawn immediately above it on the chart. The other views — of the Blue Mountains, Hellshire, Salt Pond Hill, Yallahs, and Fort Augusta — were carefully selected and labeled so as to guide the inbound mariner safely to his anchorage. When combined with the scores of houses, plantations, visually prominent buildings, and remarkable natural features meticulously plotted on the chart, these drawings provided a truly three-dimensional guide for the navigator. One who studied and mastered Gauld's "General Plan" could proceed where he pleased with the greatest of confidence in any part of the waters surrounding Port Royal — with a certain qualification which the surveyor duly noted. Wind and sea changed the scene constantly and could do so dramatically. Drunken-man's Kay was "said to have been once the largest of all the Kays, but was in a manner destroyed by a hurricane in 1722, and is only a Slip of Sand now." Similarly, Gauld noted abreast Hellshire Point "An Island formerly covered with Mangroves, but washed away by a Hurricane in 1751." The shoal that was once Old Port Royal was not apt to be forgotten, but the point seemed to have a life of its own. The eastern wall of the town was "built as a defense against the Sea, which used to beat against it, and also the Fort; but at present the beach is at a considerable distance from both, and the Point increases daily." The building process (which continues today) was also evidenced by the terraced "Traces of the different Beaches" sketched by Gauld at the end of Port Royal Point.

Gauld's two "General Plans" differ considerably in minute details but not in any important manner. Both are handsome examples of the surveyor's craft and the draftsman's art, and they must have cost Gauld many agonizing weeks bent over his drawing board. They declare his mastery of every facet of his trade, and in one corner they display the proof of his honor and generosity, for he added the note "In taking this Survey I have been very much assisted by Lieut. Charles Knatchbull & Mr. John Payne." That "N. B." served as a gracious farewell to Knatchbull and "welcome aboard" to a remarkable seaman with whom George Gauld would survey many more islands.

Admiral George Rodney made clear his personal interest in the naval surveyor's work on February 18, 1773, when he paid a visit to the *Earl of Northampton* and allowed Lieutenant Phillips to carry him from Port Royal across the bay to Greenwich.[7] Rodney was a stickler for service etiquette, and young Phillips would have seen to it that the admiral received full honors while aboard his ship. George Gauld doubtless accompanied the party, and it is easy to imagine the surveyor using the opportunity to demonstrate to Rodney the efficacy of his latest chart. There was evidently considerable relaxation aboard *Northampton* when the commander in chief was safely ashore — and even greater embarrassment when, returning to Port Royal after dark, the sloop went aground on a mud bank in the narrows, just a cable's length from the northern buoy! Phillips's navigation had been thrown off by an ebbing tide; he got his ship off the bank with the incoming tide, but the captain of the surveying sloop must have returned to Port Royal a bit crestfallen.

Rodney's support of George Gauld's endeavors took a more tangible form five days later. To assist the cartographer the admiral sent to *Northampton* a new surveying boat, one large enough to mount two masts with sails, hence sometimes referred to as a schooner by Phillips, who thereby elevated himself (though only in his own mind) to the rank of commodore of a tiny squadron. Unofficially christened *Florida* by the surveying team, this little craft would prove herself highly seaworthy, perform yeoman service for Gauld during the summer of 1773, and enable him to extend significantly the scope of his observations among the keys and islands of the Gulf of Mexico.

The working season was fast approaching, and during the first two weeks of March, *Northampton* was taken for a shakedown cruise to Cape Tiburon, the southwest point of Hispaniola. The experience indicated that both ship and crew needed further preparation before they undertook an extended voyage. Carpenters from the navy yard got to work on the sloop at once; the survey boat *Florida* was also altered and fitted early in April, and on the seventeenth a second, smaller boat was sent to *Northampton* for Gauld's use. Both boats and their mother ship were ready by April 24, and three days later Lieutenant Phillips made sail, leaving Port Royal on a westerly course, the small boat in tow, the little schooner proudly sailing along in company. The admiral's blessing went with them, for promptly upon leaving harbor, Lieutenant Phillips put the ship's company on "⅔rds allowance of provisions of all [kinds] . . . Except Rum, by Order of

7. Captain's log, *Northampton:* ADM 51/4178.

Jamaica to Key West, 1772–1773

Sir. Geo. Rodney." Such attention to detail would have been surprising had not Rodney been fanatically devoted to the virtues of turtle meat as a dietary supplement for his sailors! Phillips had the admiral's orders to flesh out his men's rations with that special delicacy of the islands, and *Northampton*'s crew would indeed spend much of their time, while the surveyor was off on his business, seining fish and trapping turtles.

Rounding the western end of Jamaica, *Northampton*'s course lay northwest toward Grand Cayman Island.[8] Lieutenant Phillips found that he had to shorten sail to allow the surveying schooner to keep up with him, but on May 4 the two vessels reached their first destination and anchored in a little cove at the west end of Grand Cayman, picturesquely known as Hogstie Bay. No sooner was the sloop secured than Gauld and his surveying team were in their boats and away to Sandy Key, the southwest tip of the island, to begin their observations. The next day Gauld began a circumnavigation of Grand Cayman that would occupy him for nine days and enable him to chart the whole of its reef-bound coast in detail.

The surveyor's interest in this lonely twenty-eight-mile-long island was set forth in the "Remarks" appended to his finished chart of Grand Cayman:

This Island, being in the track of all the homeward bound Jamaica Vessels, on the[i]r passage round the West end of Cuba, & through the Gulf of Florida, is frequently the cause of shipwrecks. It is low . . . & very dangerous in the night; being surrounded by reefs, beyond which the soundings do not generally extend more than a quarter of a mile. The only place of Anchorage is at the West end.[9]

The dangers of Grand Cayman were displayed plainly enough on Gauld's chart. On the east end of the island, at Gun Bluff, he located the spot where the transport *Cumberland*, carrying troops for Pensacola, had been lost in 1767; on the northern reef the *Augustus Caesar* had been "cast away" in 1765, and the debris of yet another "old wreck" was to be seen. At no point save Hogstie Bay was the shore accessible to seagoing vessels; even the seven-mile-wide mouth of North Sound was closed to all but small craft. The utility of his new boats was evident as Gauld worked his way around the island, sounding inside the reef, noting the island's few distinguishing fea-

8. Ibid.
9. "The Island of Grand Cayman. Surveyed by Geo. Gauld 1773": MODHD, 196 Ag 1.

tures and its several plantations, and tracing the course the sea had taken in 1751 when "a remarkable hurricane" swept waves right over the narrow southern dogleg of Grand Cayman, temporarily cutting a channel from Little Pedro Point into North Sound.[10]

By May 16, Lieutenant Phillips was becoming nervous about the waves — "a prodigious swell" — rolling from the west into Hogstie Bay, and as Gauld had finished his work, *Northampton* and her little convoy prepared to sail for Cape Antonio, the westernmost tip of Cuba. On the twenty-second the sloop lost sight of the little *Florida* but fired a swivel gun and, hearing a musket shot in answer, regained contact. On May 23, Phillips anchored his ship off Cape Antonio and remained there until the end of the month. Gauld was away from *Northampton* much of this time, and having sounded around the cape and its bay, he extended his investigations eastward among the islets called the Coloradoes just off the Cuban mainland. Phillips noted considerable maritime traffic around Cape Antonio and must have recalled the fate of his predecessor on *Northampton*, Lieutenant Gibbs, when a Spanish *guarda costa*, escorting several vessels to Havana as prizes, sailed past his position. Phillips was Rodney's man, however, and although he was in Spanish waters, he put a shot across the bow of a passing brig that failed to show her colors and made her prove she was a British merchantman.

The next day, May 31, Phillips pulled away from the Cuban shore to begin five frustrating days cruising northwest of Cape Antonio, vainly hunting for an uncharted rock that had been reported by the skipper of the packet running between Port Royal and Pensacola. On June 5 he gave up the search and headed north and east for the Tortugas. Proceeding most cautiously over the Tortuga Bank, both *Northampton* and *Florida* sounding as they went, the surveying expedition found an anchorage among the keys of the Dry Tortugas on June 9, and Gauld began another major cartographic project.

The Dry Tortugas, outermost and least impressive of the long chain of islands swinging west from the tip of Florida, marked the northern extremity of the straits leading from the Gulf of Mexico to the Atlantic. Little more than sand and coral banks barely rising above the sea, they had little to recommend them to the attention of anyone save the lonely mariner. To him and to his chart-maker, their location and configuration and the shoals surrounding them were vi-

---

10. Rough draft, "The Island of Grand Caymana," MODHD, U10 Ag 1.

tally important. George Gauld would spend two long months among these cartographic flyspecks, carefully plotting and sounding the distinctions of land and sea, visible and invisible, that made up this desolate place. Apparently the surveyor worked the keys from southwest to northeast, but so small was the area under study that he could easily retrace his steps at any time.

Life aboard *Northampton* was a dull routine once a comfortable anchorage was found near Middle Key. Her crew had little to do except to supply the smaller boats and keep them fit and clean. She moved, from time to time, seeking a better anchorage or a more convenient spot from which to serve the working party, and her people manned the survey boats when required. A coconut tree was planted on the Southwest Key, another on the Middle Key, as marks for the surveyors; the absence of water discouraged much hope of their growing, but the expedition also planted potatoes and yams, Guinea and Indian corn, and peas, either anticipating a return some day or as a forlorn hope for castaway seamen on that barren strand. The islands did attract some mariners. On June 19, the expedition welcomed a schooner "from Providence here aTurtling," and the visitor remained until July 5. There was also the problem of maintaining a supply of drinking water. Barrels were sunk in the sand of North Key in the hope of collecting water, and some was secured by that means, but early in July it proved necessary to dispatch the little *Florida* to Key West, where the water casks could be filled more readily, and to repeat the trip later in the month. Lieutenant Phillips diligently pursued the great sea turtles that nested on the islands and found that they were quite capable of destroying the nets with which he sought to capture them. Several loggerheads and at least one "fine large Green Turtle" were caught and duly logged as eaten, thereby satisfying Admiral Rodney's admonitions. Phillips did not record the reaction of *Northampton*'s crew to this local delicacy, but they remained remarkably healthy on their prescribed diet.

The months of June and July were generally fair, but about mid-June the winds began to rise; the smaller boat was hoisted aboard *Northampton,* and *Florida,* in which George Gauld had just departed for a six-day surveying run, hurried back to the Middle Key anchorage to ride out the storm beside her consort. The squalls increased; heavy seas built up; on June 13, Phillips reported "the Sloop rowling Gunwale in, & Pitching very much. At 10 AM the Florida parted from her Morrings, sent the Boat & got her up." The next day the storm worsened. With *Northampton* "labouring very much," the best bower parted, and, having lost her anchor, the sloop drove upon a coral bank. Fortunately she was not damaged, but her men were to spend many weary days searching for the lost anchor.

As both surveying boats were in almost constant use, it is evident that Gauld worked two teams simultaneously, one under his own direction, the other headed by John Payne. They were usually away from the sloop for six days at a time, and on one occasion Gauld was absent for two weeks, and that in the smaller boat, for the surveyor worked from either the schooner *Florida* or his smaller craft as the circumstances of their seaworthiness dictated. At roughly three-week intervals it was necessary to carry out minor repairs on both boats, to pay their bottoms with turpentine and caulk their planking.

By August 7, the surveying of the Dry Tortugas was completed, and *Northampton* worked her way into open water and began searching for a bank presumed to lie off the Southwest Key. After five days Gauld apparently gave up, and the sloop headed east, anchoring at Key West on August 16. There the surveyor would establish the co-ordinates for the Florida end of the long chain of islands extending into the Gulf of Mexico. On August 20, he and Lieutenant Phillips went to the southwestern end of Key West to take simultaneous observations at sea level at high noon. They made their computations separately and by different methods. Gauld determined the longitude to be 77°31′; Phillips recorded 77°32′. The naval surveyor, taking the declination from "the Ephemeries" or astronomical almanac, found the latitude to be 24°29′ N, while his ship's captain got 24°32′ N by using a mariner's compass.[11]

Key West offered protection for the sloop and convenient landing places for Gauld's working parties. As the schooner was undergoing repairs, the surveyor stocked the smaller boat for fifteen days and set out for an extended exploration on August 24, only to reappear on the twenty-seventh to report that he had accidentally broken the glass of his mariner's compass and could not proceed further without it. Lieutenant Phillips supplied him with another compass, spare

---

11. Captain's log, *Northampton:* ADM 51/4178. The coordinates of the modern naval station, near which Gauld and Phillips must have taken their readings, are 81°48′ and 24°34′, respectively. Gauld subsequently charted the southwest corner of Key West at about 24°29′ and 82°35′, variation 7° E. See "A Plan of the Tortugas and Part of the Florida Kays; Surveyed by George Gauld M.A. 1773. For the Right Honourable The Board of Admiralty." MODHD, D966 88.

glass, and card, and Gauld resumed his work. He returned to *Northampton* September 8, and was gone once more from the twenty-second to the thirtieth of September. When *Florida* was ready for use again, Phillips sent her to Sara Gold Key (now Boca Grande Key) for water.

Nathaniel Phillips found Key West more interesting than the Dry Tortugas, for it was a watering point for considerable shipping. When *Northampton* first arrived, she encountered a Spanish schooner from Havana whose skipper offered the dubious explanation that he was waiting to receive a shipment of money to be transported to Jamaica. Four more Spanish schooners appeared in September, at least two of them bound for the fishing grounds off the coast of the Florida peninsula, and two British vessels from Providence put in at the end of the month. Just as George Gauld had had problems with equipment, so John Payne, his assistant, ran into difficulties and had to replace "one half hour glass in lieu of one lost overboard." Occasional loss of material was not surprising, of course; the important fact was that Gauld had carefully stocked the tender so that any losses could be made good on the spot.

After a month and a half at Key West, Gauld was satisfied with the season's work. He wished to fill in the general outline of the sandy keys sweeping away to the west, so the expedition began to retrace its course in order to establish more accurately the locations of the islands between Key West and the Dry Tortugas. The small surveying boat, with fifteen days' rations, left Key West October 3, bound for the Marquesas and then the Tortugas, where she would await the sloop. Phillips sailed on October 6, sending the schooner ahead to Marquesas Keys and himself spending three days examining the banks and reefs just off Key West — which may suggest that George Gauld stayed aboard *Northampton*. On the ninth they picked up *Florida* at Marquesa, and both vessels anchored the next day, along with the smaller boat, north of Bushy Key in the Dry Tortugas.

It was well they arrived when they did. October 10 began with squalls and ended in conditions which Lieutenant Phillips, with notable understatement, described as "very Tempestuous." When the wind rose, *Northampton* began to drive toward a bank of coral, and having no time to weigh anchor, Phillips cut her cable and fought the wind to what he hoped would be a safer haven southwest of Bushy Key. There the ship grounded on a sand bank. Having lost his anchor, Phillips lashed five swivel guns together to make an improvised grapnel with which to draw the sloop into deeper water; by nightfall he was afloat and pleased to find that no structural damage had been done to the vessel. Not until the thirteenth could any repair work be done, so persistent were the squalls, and taking into consideration the little surveying boats for which he was responsible, it was the sixteenth before Phillips dared to work his way out to the open sea.

The battered flotilla headed north for Pensacola, but late October weather was so bad that they had little idea what their track was or where they might make the coast. On the twenty-second Phillips encountered and halted the sloop *Sally*, bound for Pensacola from New York, whose master believed himself to be somewhere west of Cape San Blas. Together *Northampton* and *Sally* bucked increasingly heavy seas, discovered that they were directly off the cape, and barely managed to claw their way into the shelter of St. Joseph Bay. There they waited out the storm for forty-eight hours while Gauld sent men ashore to cut spars for future use as standards for his surveying flags. At last, on the twenty-sixth, they began working *Northampton* over the bar. Abruptly the wind shifted; the sloop missed stays, fell off, and, drifting, struck bottom two or three times. Phillips hauled her off in good order and recorded his relief at finding no water in the pumps, an indication that he had again escaped major damage. Proceeding slowly on to Pensacola, *Northampton* moored there at the end of the month. After an absence of two years George Gauld had returned home.

Lieutenant Nathaniel Phillips took only enough time to replenish his supplies and check the condition of his mast before he sailed for Jamaica on November 9. He left the two surveying boats behind, in the care of the boatswain and carpenter of H. M. S. *Ferret*, to be beached and thoroughly reconditioned before the next season. Phillips and *Northampton* had had their fill of surveying. They reached Port Royal on Christmas Day, worn out by an exceptionally rough return voyage, and on the last day of 1773, a formal survey was conducted by the Master Shipwright who found the sloop "entirely unfit for further Service."[12] Men and stores were distributed in every direction, and on January 18, 1774, Lieutenant Phillips officially discharged the surveyor and his servant, Peter Quash, as well as him-

12. Captain's log, *Northampton:* ADM 51/4178; lieutenant's log, *Ferret:* ADM L/F/70A (NMM); Rodney to Admiralty, January 4, 1774: ADM 1/239.

153

Jamaica to Key West, 1772-1773

self, and put *Northampton* out of commission preparatory to her being sold. After five years, the surveying sloop deserved to be put up, but George Gauld could not be left dangling. He was placed on the supernumerary list of H. M. S. *Portland*, Rodney's flagship;[13] his pay would continue, but he must once more press from the far side of the Gulf of Mexico for a proper ship with which to pursue his work.

The proof of his dedication to the duties of his appointment was offered to Admiral Rodney in a letter completed just before *Northampton* left Pensacola. Gauld outlined his recent activity and offered his own evaluation of its importance, but he felt obliged to point out to Rodney that "it is only about a week since we arrived here, having been out six months and a few days from Jamaica." It would, therefore, be some little time before he could "send fair copies of our summer's work." For the present he explained that:

As the island of Grand Cayman lyes directly in the way between the West End of Jamaica and Cape Antonio, and has occasioned many shipwrecks, we took a Plan of it sufficiently exact to be of service to His Majesty's Ships and others that pass that way. Afterwards we stopped at Cape Antonio, took a Plan for several miles on each side of it, with part of the Colorades, and then went in search of the shoal which Capt. Clarke of the Diligence Packet imagined he saw bearing about NW 10 or 11 Leagues from Cape Antonio. We traversed for several days for it but to no purpose; we could see nothing of it, and therefore bore away for the Dry Tortugas.

We finished them and continued the Survey among the Florida Keys as far as Cayo Huesso, commonly called Key West about 22 Leagues to the Eastward of the Tortugas.

The surveyor assured Rodney that his limitation of "two thirds allowance [of victuals] was plenty of Provisions, especially as we found great quantities of Fish and Turtle."

Gauld's immediate concern was, of course, the replacement of *Northampton*. "In case another vessel should be purchased for this Service, I would beg leave to mention some qualities she ought to have. In the first place she should not draw above 6½ or 7 ft. at the utmost when deep. She ought to be strong, roomy and commodious, and a little larger than the Northampton." He hoped that the new ship might be dispatched in sufficient time to reach Pensacola by March or early April in order "that we may have the whole Summer

13. Muster tables, *Earl of Northampton* and *Portland:* ADM 36/8519 and 8047. John Payne was discharged on January 17, 1774, like Gauld, to *Portland*.

before us." Gauld evidently planned to return to the Florida cape, for he asked the admiral's permission to secure provisions from naval vessels at Providence, in the Bahamas, if the need should arise.

Referring to his previous year's work at Jamaica, the surveyor hinted strongly, "I should be glad to know whether their Lordships [at the Admiralty] will permit the Plan of Port Royal Harbour to be published." At the same time he called to Rodney's attention the competence and the contributions to surveying made by John Payne, who had returned to Jamaica aboard *Northampton*. He hoped his assistant might receive supplementary wages from the Admiralty and assured Rodney that should he "want the rest of the Buoys to be placed [in Port Royal harbour], Mr. Payne can do it as well as if I was upon the spot. I wish that was done, because in the Directions I mentioned them as actually placed." Obviously pressed for time as the sloop prepared to sail, Gauld bethought himself of one last request: "I must beg some more large Drawing Paper if you can spare me any."[14] Having completed his annual report and having said farewell to the *Earl of Northampton*, Gauld settled down to a winter in Pensacola and the drafting of his latest charts.

The plan of "The Island of Grand Cayman," a modest effort, useful though it might be to seafarers, requires no further comment. George Gauld's major production for 1773 was "A Plan of the Tortugas and Part of the Florida Kays," a great chart seven and one-half by two and one-half feet in size, whose scope extended from the western end of the Tortuga Bank to Key West, a straight line distance of roughly eighty miles. The soundings which Gauld recorded on either side of the central keys indicate that he actually worked a good ten miles north and south of this axis, thereby surveying some 1,600 square miles of ocean, reef, and island — a tremendous accomplishment on the part of the surveyor and his team. In addition to the usual indications of landmass and sea depth, the chart included panoramic views, taken from several distances and directions, intended to enable the navigator to recognize the Marquesas Keys, Sara Gold (Boca Grande) Key, Bluff and Double (Man and Woman) keys, and Key West. The Dry Tortugas simply offered no feature capable of reproduction. At the bottom of his chart Gauld appended lengthy "Remarks" of an explanatory and descriptive nature no less interesting than the draft itself.

14. Gauld to Rodney, November 8, 1773: ADM 1/239.

Jamaica to Key West, 1772-1773

Addressing himself first to the Dry Tortugas, Gauld observed:

A thorough knowledge of the Tortugas, and that extensive Bank near the extremity of which they are situated, is very essential both for the navigation of the Bay of Mexico and the Gulph of Florida, as they form an elbow between the two, at the distance of about thirty Leagues from the nearest part of Florida, forty from the Island of Cuba, and 14 leagues from the westermost of the Florida Kays. The Dry Tortugas . . . are ten in number: most of them are covered with small bushes, and they may be seen about 4 leagues distance. The South-westermost, which is one of the smallest of them, lyes in 24°, 32', 30" North Latitude and about 83°, 45' West Longitude from the Royal Observatory at Greenwich.

Though the Tortugas are generally looked upon to be very dangerous, and to a person unacquainted with them, undoubtedly they are so, especially in the night; yet when known, they may be found on many occasions to be very useful and convenient. There is good Anchorage in several places, particularly in a small but snug Harbour at Bush Kay, which is entirely sheltered from the Sea by a large Reef of Rocks; and a flat shoal within them about half a mile broad. The bottom is soft Clay and Mud, and this Harbour is quite smooth even in a gale of wind.

Although his longitudinal calculation was dangerously incorrect by modern standards, Gauld's measurement of the latitude was extremely accurate.[15] The physical appearance of the keys has changed little in two hundred years, save for the erection of Fort Jefferson in the mid-nineteenth century, but their number has declined as hurricane after hurricane has swept over them. The surveyor and Lieutenant Phillips noted the effect of storms in 1773, and in 1935, a great hurricane destroyed Bird Key, reducing the number of dry keys to a half dozen. The shoal configuration has changed only slightly since Gauld charted the Tortugas, and the channels entering the area remain the same. Loggerhead-Turtle and East Kay retain their names, but modern Hospital and Garden keys reflect the impact of human occupation upon islands Gauld designated Middle, Booby, Bush, and Rocky Kay.

There is no drinkable water to be got on any of the Tortugas except the Northermost, and even that is very brakish: nor is there any firewood

---

15. The following comparisons are based upon C. & G.S. 1351, *Sombrero Key to Dry Tortugas*; AMS 4433 3 NE-Series V847, *Dry Tortugas*; AMS 4533 2 SE and SW-Series V847, *Marquesas Keys*.

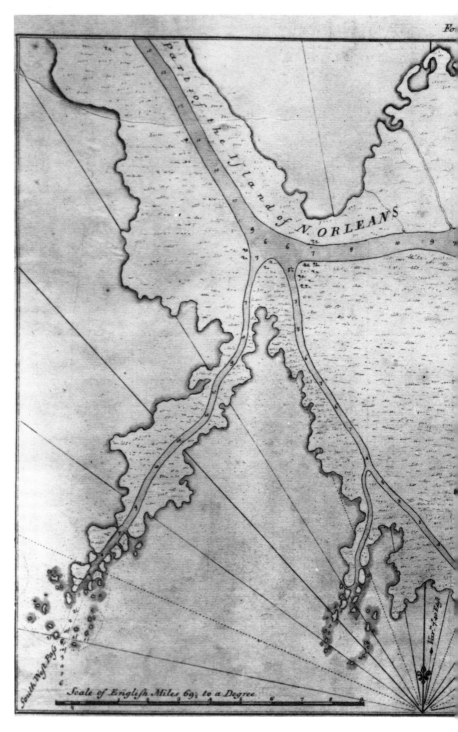

X. "A Plan of the Mouths of the Mississippi." Gauld repeatedly surveyed the entrances of the river, noting changes in the depth of water on the bars and in the passes. This chart was the basis for his suggestion that Great Britain

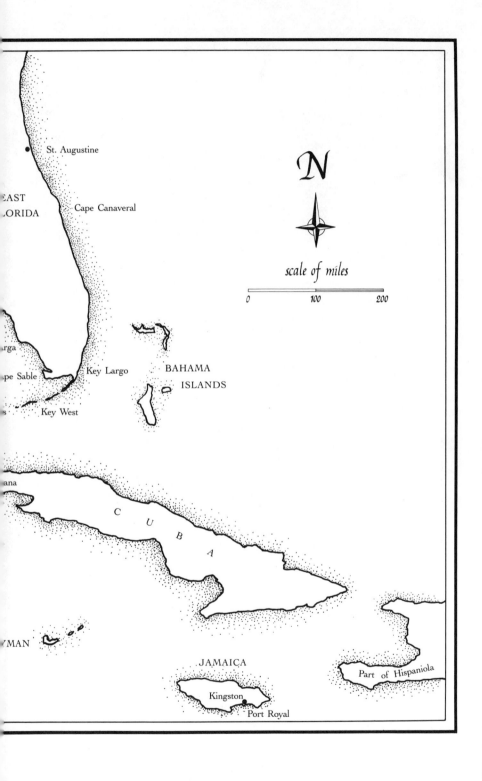

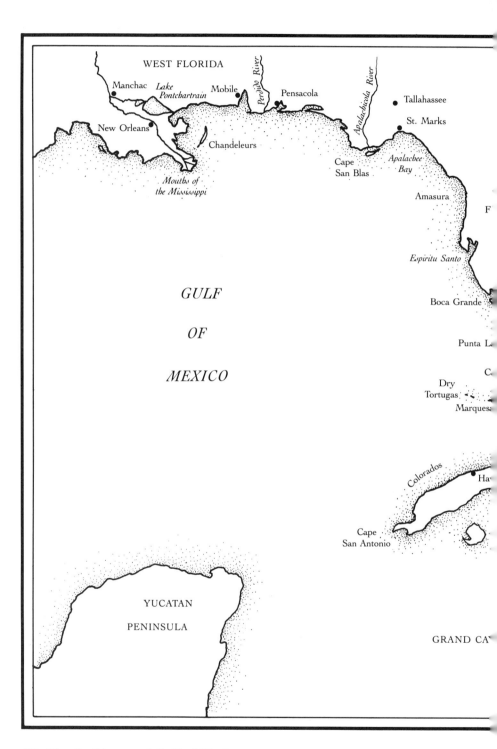

XI. The Caribbean and Gulf of Mexico

should claim the easternmost pass as the proper boundary between West Florida and Louisiana. (Courtesy of the Public Record Office.)

except a few bushes which it were a pity to cut down, as they serve to distinguish the kays at a distance.

The Tortugas abound with variety of Sea-birds, Turtle, and excellent fish.

It is delightful to discover that the eighteenth-century surveyor was concerned with environmental protection in an area that is today dedicated to that cause. His concern was, of course, that of a seaman; it is rather the great nestings of terns and turtles that attract twentieth-century naturalists to the Tortugas. The very name, attached to the keys by Ponce de León in 1513, denotes the swarming presence of the huge turtles that once populated the islands before a growing Cuban appetite for their meat and eggs threatened their existence. Neither the Spaniard nor the Englishman, however, realized that these shoal waters were one of the world's greatest shrimping grounds.

Reverting to the navigational problems the Tortugas posed, Gauld advised mariners to sound frequently at night when between 24° and 25° latitudes and to avoid anything less than thirty fathoms. "The Spanish Ships from Vera Cruz bound to the Havanna," he wrote, "make it a constant practice . . . to strike soundings on this Bank" and to anchor regularly at dusk, "but there is no necessity for so much caution." Certainly, with George Gauld's chart, an English mariner could proceed more boldly than his rivals.

Gauld advised that there was "deep water and a safe and broad Channel for the largest Ships" east of the Tortugas, but he warned of the dangers of the coral bank now called Rebecca Shoal and the Quicksands farther east. Cayo Marques, "the Westermost of the Florida Kays," now Marquesas Keys, deserved little comment. "The Trees on this and some of the contiguous Islands are thick and pretty high, and may be seen at a considerable distance." He designated the large island Cayo de Boca Grande, and of the waters encompassed by its curving shore he wrote, "This large Bay is quite shoal: so that even a Canoe can hardly find a passage at low water." Comparison of Gauld's "Plan" and modern charts suggests that like the Tortugas, the Marquesas Keys have changed considerably since the eighteenth century because of oceanic erosion.

Next came the Boca Grande Channel, "a large Opening [in the Florida Reef] to the Eastward of Marques, but it is a very indifferent Channel for any thing but small vessels." The next key, which today carries the name Boca Grande, was designated by Gauld as Sara Gold Kay; his Double Kay and Bluff Kay are now known as Woman and Man keys. The smaller keys to the north were precisely located but left unnamed, but modern Crawfish Key was even then Crayfish Kay.

And so George Gauld came to "Cayo Huesso, (or Bone Island,) corruptly called Key West." Offering fresh water and good anchorage, "it is much frequented by the Turtlers and Wreckers from Providence, and Spanish Vessels from Cuba, in their way to and from the Coast of Florida, where they are annually employed in catching and curing Fish for the consumption of the Havanna, and other parts of the Spanish Dominions in the West Indies." "It is full of Lagoons and Swamps," Gauld wrote, and as he wandered among the ironwood, fustic, and mangrove trees, he noted the presence of the little Key deer, no bigger than a dog.

In the course of describing the approaches to Key West, Gauld remarked that "in several places the large branches of Coral rise above the water." Regular three and one-half foot tides from the south and southeast were measured to the north of the reef, but on its southern side "the Current sets almost constantly to the Eastward, till it meets with the Bahama Bank, which gives it a turn to the Northward through the Gulf of Florida, where it runs with uncommon rapidity, till it mixes with the Waters of the Atlantick Ocean." Having raised the topic of the Gulf Stream, however, Gauld dropped it: "the discussion of this important Subject must be deferred for the present, till further Observations shall afford proper and sufficient Materials."

Now that he had charted one extremity of the Florida Keys on the "Plan" of 1773, George Gauld was ready and no doubt eager to trace the islands stretching east behind Key West. That task was scheduled for the coming season, as Gauld perfected his charts during the winter of 1773-74, but he must await a ship; before it arrived he would seize the opportunity to return to the western limits of the British Floridas.

CHAPTER X

Exploring the Limits and Beyond, 1774-1778

Admiral George Rodney's appreciation of the work of the naval surveyor was demonstrated by his immediate attention to Gauld's plea for assistance. The *Earl of Northampton* anchored at Port Royal on December 25, 1773; in less than a fortnight Rodney had her surveyed and advised the Admiralty that she would be sold at public auction. The storekeeper at Port Royal was directed "to be very Attentive in the purchase of another Surveying Vessel by complying with the desire of Mr. George Gauld in her draft of Water and Demensions, taking care to put the Government to as little expence as possible in the said Purchase." The admiral assured Their Lordships that he would "hasten the said Vessel to Mr. Gauld as early in the Spring as he desired"; he anticipated "great Progress . . . in Surveying the Coast near Cape Florida" during the summer of 1774.[1] Two weeks later *Northampton* was decommissioned, and by the middle of March, Rodney could report that the storekeeper, "with the assistance of Mr. Beall, Master Shipwright of this Yard," had located "a new sloop remarkably well found, and of the Dimensions expressed in Mr. Gauld's letter." On their recommendation Rodney purchased the vessel for £670, named her the *Florida,* and appointed Lieutenant Charles Cobb to command her. The admiral hoped to dispatch her to Pensacola in March, but both he and the surveyor were to be disappointed in their expectations of an early start that year.[2]

After an absence of nearly two years, George Gauld would have found many changes at Pensacola. Governor Peter Chester brought a degree of stability to West Florida politics that was enhanced by greater prosperity, itself owing something to the partial restoration of the military presence during the Anglo-Spanish tensions of 1770-71. The governor allied himself with the merchants of Pensacola and enjoyed support from the government at home, and his West Floridians seemed to lose interest in politics. The General Assembly of 1771 quietly passed both necessary and useful legislation and was

1. Rodney to Admiralty, January 4, 1774: ADM 1/239.
2. Rodney to Admiralty, March 14, 1774: ADM 1/239. *Florida* left Port Royal between April 23 and 30: ADM 36/7978.

159

dismissed. In 1772, however, Governor Chester's insistence upon maintaining control of the electoral process collided with the determination of the electors of Mobile to restrict the terms of their representatives. The resulting stalemate was broken when the governor simply dissolved the assembly before it ever met. Such was the strength of the royal governor in West Florida that it made little difference to anyone. Chester's administration, managed in large part by his secretary, Phillip Livingston, proceeded smoothly. Lieutenant Governor Durnford supported his superior in the council, although they subsequently fell out over the construction costs of a new governor's palace which Durnford designed for Chester. The most significant changes in the colony were taking place on the Mississippi River, where rich virgin lands were being taken up by settlers in such numbers as to suggest the desirability of creating new towns, even perhaps a new and separate colony. The real and speculative opportunities in the west attracted much attention, not the least that of Governor Chester.

The development of the Mississippi region depended upon several factors, the first being easy access from other parts of British America. While settlers might float down the Ohio and Mississippi rivers from the backcountry of Pennsylvania or trudge overland through Indian country from Georgia and the Carolinas, the seas offered the easiest route to these new lands; but this route was guarded, if not closed, by Spanish occupation of New Orleans. Although British subjects enjoyed free use of the river by treaty, exercise of that right depended upon the reliability of pilotage at the mouth of the Mississippi. George Gauld had noted the fact and urged the establishment of a British claim to one of the entrances to the river in 1770.

An alternative route to the Mississippi existed on paper and had long intrigued colonial authorities, surveyors, and chartmarkers. The Treaty of Paris had given to West Florida the Iberville River, which was believed to link the waters of Lake Maurepas with those of the Mississippi, thereby creating the amorphous Isle of Orleans. From their earliest appearance on the Mississippi in 1764, British army engineers such as Philip Pittman, James Campbell, and Archibald Robertson had surveyed, investigated, and attempted to clear the Iberville for navigational purposes. Their efforts had established the fact that only when the Mississippi was at flood stage did its waters ever flow through the Iberville to the lakes. In normal or dry seasons the junction of the Iberville with the Mississippi was as much as twelve feet above the great river. Further, the hopes of the army engineers that the floodwaters of the Mississippi would sweep the Iberville clear of logs and debris were totally misplaced. Each year the channel of the Iberville was clogged once more with the refuse left by spring floods, and its upper reaches were baked by summer suns. Occasionally, however, a canoe, and in 1768 a "schooner" reported as drawing four feet of water, fortuitously made its way from Lake Maurepas to Fort Bute, the British post at Manchac, thereby keeping dreams alive and leading the engineers, notably Elias Durnford, to propose the digging of a canal to accomplish what nature refused to do.[3] George Gauld's first opinion on this issue, expressed in 1769 and evidently based upon hearsay, was that "It would certainly require some trouble and expense to clear the Ibberville of logs and cut a deeper channel between it and the Mississippi, but it's undoubtedly practicable."[4] That favorable view, based upon the self-justifying reports of other observers, recommended him to Governor Chester when he considered the future of the Mississippi.

In addition to the military engineers who worked up and down the Iberville, an almost constant stream of provincial luminaries investigated the scene. Lieutenant Governors Montfort Browne and Elias Durnford, the latter in his capacity as surveyor general, visited Manchac, and Thomas Hutchins traced the river's course with a professional eye. Their exceedingly optimistic reports, together with the influx of settlers on the Mississippi, persuaded Governor Chester to view the western parts of the colony in the spring of 1774. Chester reported to Lord Dartmouth, the new Secretary of State for the Colonies, that he and George Gauld were on the Mississippi in March of that year. There is no question that George Gauld, with Doctor John Lorimer, his good friend and a scientist, and Major Alexander Dickson, the ranking army officer in West Florida, did indeed make such a trip, but there is reason to think that the report that Chester submitted was drafted by someone else (it was written in the first person) and accepted by the governor without careful editing, for Chester was certainly in Pensacola in February and March when the three friends were in the west, and he was touring the Mississippi in November when George Gauld was at sea. It ap-

3. Douglas S. Brown, "The Iberville Canal Project," *Mississippi Valley Historical Review* 32 (1946): 491–516; Robert R. Rea, Introduction to Pittman, *The Present State of the European Settlements on the Mississippi*, pp. xxiii–xxv.

4. Gauld, "A General Description of West Florida, 1769," pp. 6–7.

pears that Gauld, Lorimer, and Dickson left Pensacola together on February 16; they visited Manchac, then pushed upriver to Natchez where Dickson left the party. Gauld and Lorimer went on to the Yazoo, the northwestern corner of British West Florida, a point of cartographic interest to the surveyor. As they proceeded, Dr. Lorimer contributed to Gauld's work by "taking the latitude" at Natchez and other geographically prominent sites. Major Dickson was back in Pensacola by April 30; his traveling companions probably returned early in May.[5]

Whatever their itinerary and timetable, Gauld and Lorimer took a long look at the River Iberville and at the conjunction of the Mississippi and the Houma River, which appeared to be a more promising corridor of communication than the Iberville. Of the mouths of the Houma, Chester reported to Lord Dartmouth, "When Mr. Gauld and I passed them, going up the Mississippi in March 1774 [they] had full 13 feet water tho' at that time there was not above as many Inches in the Ibberville at Manchack."[6] Although the governor's report stated that the French and Indians "go through there in their Canoes to Lake Maurepas," there was clearly little to be hoped for in the way of effective commercial communication unless the home government was prepared to expend huge sums for canal building. Never likely, that prospect disappeared with political developments north of the Gulf Coast in 1775.

For George Gauld, the trip up the Mississippi was an unexpected opportunity to visit and chart "the extreme Barrier of His Majesty's Dominions in North America." Recognizing that he had "rather gone beyond my Limits as a Sea Surveyor," Gauld justified himself by the fact that "all above New Orleans was taken on a private Party, at a season when I could not go on the general Survey of the Coast, at our own expence without costing Government anything."[7] The journey gave Gauld the chance to draft a plan of Fort Bute and to make his own observations of the Mississippi. He and Lorimer were undoubtedly curious about the western lands which both had secured in recent years. The initiative must have been provided by Dr. Lori-

5. Dickson to Haldimand, February 15, May 9, 1774: Haldimand Transcripts; Dickson to Gage, November 12, 1774: Gage Papers; *The South-Carolina and American General Gazette*, vol. 17, no. 825 (July 1–8, 1774), pp. 188–89.

6. Chester to Dartmouth, January 23, 1775: C.O. 5/592. The wording of this report suggests that it may have been compiled by John Lorimer.

7. Gauld to Admiral Peter Parker, January 14, 1779: ADM 1/241.

mer, who held a thousand acres on Thompson's Creek and a second grant north of Natchez. As Lorimer had corresponded with Gauld while the surveyor was in Jamaica, it seems reasonable to assume that he was responsible for Gauld's own petition of February 11, 1772, for "a Grant of one Thousand Acres on purchase Situated on Thompson Creek next to Lands advised to be Granted to John Lorimer." It was passed on July 28 and £5 was paid, the purchase rate being five shillings for each fifty acres.[8] In June 1773, Gauld entered into private negotiations for the acquisition of a further thousand acres on the east side of Thompson's Creek. Judging by the number of prominent businessmen and provincial councilors who also bought land in this region opposite Point Coupée, it was considered to be a good investment.

The trip enabled George Gauld to draft "A Plan of Manchac" that would become the definitive description of that ill-founded and ill-fated little British post on the Mississippi. In addition, he and Lorimer ran a line to determine the drop from the Mississippi to the juncture of the projected Iberville canal, and on April 3, when its floodwaters were only eleven inches from the top of the bank at Manchac, Gauld spent some time sounding the great river.[9] It is difficult to assess the importance of Gauld's charting of the Mississippi River above Manchac, for he had access to the earlier work of Pittman and Hutchins (who had drawn heavily upon Pittman's chart), and the West Florida surveyors often exchanged information and observations with considerable freedom. In any case Gauld's version of the Mississippi boundary would not be incorporated in a major chart of British West Florida until 1779, by which time his work had come to an end—and the river had been lost to Spain.[10]

For the moment, Governor Peter Chester's interest centered upon securing more dependable access to the mouth of the Mississippi, and in that pursuit he was enthusiastically abetted by the naval surveyor. There can be no question that Gauld persuaded Chester to accept the argument he had offered in 1770 that "those Islands which are called the New Balise, belong to the Crown of Great Britain, and

---

8. "A State of all Grants of Land" in Chester to Dartmouth, December 20, 1773: C.O. 5/590; Council minutes, C.O. 5/629 and 634; Indentures of June 21, 22, 1773: C.O. 5/612: 336–40.

9. "A Plan of Manchac" in the Clinton Papers, William L. Clements Library Map Collection, Ann Arbor, Mich.

10. Gauld to Parker, January 14, 1779: ADM 1/241.

Exploring the Limits and Beyond, 1774–1778

that His Majesty has a right to them by the treaty of Paris." When reporting to the colonial secretary on "the reputed Boundaries of the Province," Chester repeated verbatim Gauld's analysis of the British claim and referred the secretary to the surveyor's chart of the mouths of the Mississippi.[11] It is unclear to what extent the governor was responding to the desires of English merchants and settlers up the river or was himself seeking to initiate events. He reported to Lord Dartmouth that "in the beginning of 1774 . . . John Salkeld set himself down upon . . . Isle Verte, as Pilot for the British Vessels trading to the Mississippi, and was recommended by the Principal Merchants concerned in that Commerce as a proper Person for that purpose." Indeed, on April 20, hot on the heels of the surveyor's trip, a petition in the pilot's behalf was submitted to Chester from a number of British merchants. On May 16, Chester issued a permit for Salkeld to carry out such duties, and in June he justified his action by reference to Gauld's chart.[12] Lord Dartmouth was readily convinced that a British pilot's services were necessary, but he was not prepared to approve of Chester's proceedings without knowing the exact spot upon which Salkeld was settled, for he expected that, wherever it might be, the Spanish government would object.[13]

Chester defended himself at some length, again citing "Mr. Gauld, who is throughly acquainted with the Geography of this Country," and declaring that Spanish Governor Unzaga acquiesced in the English pilot's presence and employment on Green Island. Chester hastened to assure the secretary, however, that he had no intention of disrupting international relations and had not given Salkeld "power to take possession of those Islands."[14] There the matter rested in 1775. Lord Dartmouth's attention was focused upon Boston rather than the backwaters of West Florida, but at least one of the merchants who had expressed support for this not very subtle encroachment upon Spanish sovereignty in Louisiana, James Willing, would soon take an aggressive, if hostile, interest in the British presence on the Mississippi.

George Gauld was back in Pensacola by early May 1774, ready to

11. Chester to Dartmouth, January 23, 1775: C.O. 5/592.
12. Chester to Dartmouth, June 7, 1774 and enclosures: C.O. 5/591. The tinted chart forwarded by Chester, "A Plan of the Mouths of the Mississippi By George Gauld M.A." and designated "For the Right Honourable The Earl of Dartmouth," is in P.R.O.: M.P.G530.
13. Dartmouth to Chester, October 5, 1774: C.O. 5/591.
14. Chester to Dartmouth, January 23, 1775: C.O. 5/592.

embark upon further exploration in the sloop *Florida*. His official connection with this new surveying vessel began on February 3, 1774, when he and his servant, Peter Quash, together with his assistant, John Payne, were transferred to her supernumerary list from that of Rodney's flagship, *Portland*. *Florida* sailed from Port Royal in the last week of April and was anchored at Pensacola by the end of May. Within two weeks Lieutenant Cobb and the surveyor were at sea and making for the Dry Tortugas. In the absence of *Florida*'s logs, Gauld's activity during this and subsequent seasons cannot be followed with precision, but by the end of the first week of July he was in the Tortugas. *Florida* remained there less than two weeks before sailing east. Between July 29 and October 7 she was at Key West, and during the next four weeks she was working between the Keys and the Florida Reef. During the second week of November, Lieutenant Cobb left Bahia Honda, and the party was back at Pensacola by November 29.[15]

In his annual report to the new commander in chief at Port Royal, Admiral Clark Gayton, Gauld observed, "It was the middle of June before we could set out . . . by which means we had the worst Season of the whole year to struggle with in that disagreeable part of the world." The delay was clearly due to the failure of *Florida* to appear at Pensacola at an earlier date, a circumstance explained by the complications of readying a new ship for sea. Perhaps Gauld anticipated the problem when he went to the Mississippi with Lorimer and Dickson, but he reiterated, for Gayton's edification, "We can do more business, and more effectually, in the months of April and May, than in any other three months of the year."[16]

Florida was out on surveying duty for over five months, and Gauld's mode of operation was similar to that of the previous year. The six-oared yawl and the larger schooner-rigged boat (which could no longer be called "Florida") worked in conjunction with the sloop. "The yawl was of the greatest service in carrying on the Sur-

---

15. Muster tables of *Portland* and *Florida:* ADM 36/8047 and 7978; Gauld's notes on rough charts: MODHD D959a 88 and U.9 6g. Gauld was at Pensacola May 12 when he prepared a few "Remarks" on the Tortugas which were "left with a gentleman there" to be forwarded to and inserted in *The South-Carolina and American General Gazette*, vol. 17, no. 825 (July 1–8, 1774), pp. 188–89. These sailing instructions, essentially the same as those on "A Plan of the Tortugas . . . 1773" and later published in *An Account of The Surveys*, pp. 12–13, were made public because "it is not likely that his Draughts of the Florida Coasts will be published for some Time." It was, indeed, sixteen years before this material was printed!

16. Gauld to Gayton, December 2, 1774: ADM 1/240.

vey," Gauld reported, "as she drew little water, and in other respects was an excellent Boat for this Service." The surveyor evidently found Lieutenant Charles Cobb to be a competent commander, and he enjoyed the company of John Payne once more. To Admiral Gayton Gauld explained that:

> *Mr. Payne was Mate of His Majesty's Ship the Jamaica when she was lost upon the Coloradoes. The Summer before that accident happened he went with me on the Survey of some parts of the Coast of West Florida* [i.e., in 1769], *on which I was then employed; and he liked the business so well that he left England again on purpose to come in the Surveying Sloop, where he is of very great assistance in forwarding this Service. Sir George Rodney appointed him as Pilot, which makes it something worth his while, as he has never had any allowance before.*

As he had done earlier, Gauld suggested to Admiral Gayton that Payne, who returned to Port Royal in *Florida* at the end of the year, could be usefully employed fixing buoys upon the banks and shoals of that harbor which he and Gauld had sounded so carefully in 1772.

Inevitably, the late start in mid-June 1774 meant that *Florida* was caught far from home by the hurricane season. In spite of previous experience, on Gauld's part at least, the sloop remained in the Florida keys until early November. When she turned north toward Pensacola, tragedy struck. "Mr. Cobb will inform you," wrote the surveyor, "of the misfortune we had of losing our two Boats in a gale of wind." When *Florida* returned to Pensacola at the end of November it was assumed that both the schooner and the yawl had sunk and all hands had perished. So Gauld advised Admiral Gayton on December 2, but the next day he added a somewhat happier postscript: "Since writing the above, the Schooner Boat has arrived early this morning with the loss of two men, one drowned, and the other died with Cold and wet." Of the suffering of the survivors in that little vessel, blown by pitiless winds and tossed by cruel seas, George Gauld said nothing; the service was harsh and death never far away, as every man aboard the ships of the Royal Navy knew full well. The most useful thing the naval surveyor could add was a prayer, "I hope next year we shall be able to set out at the proper time, and to return earlier."[17]

Gauld's report reached Port Royal with the battered *Florida* on

17. Ibid.

G^eo. Gauld — Cartographer

January 2, 1775, and Clark Gayton forwarded the surveyor's observations to the Admiralty. At the same time he assured Their Lordships that he had given directions to have the sloop "Careen'd, & fitted again, as fast as possible in order to her joining him" by the first of April.[18]

The admiral was nearly as good as his word. Lieutenant Cobb left Port Royal on March 23, 1775, and *Florida* was back in Pensacola by April 23. By that date Gauld had completed a fair copy of the 1774 summer survey and sent it to Jamaica by the packet ship *Diligence*. The surveyor advised Admiral Gayton that the chart was "enclosed in a Box with a sliding Cover fixed by a Screw Nail in order that you may the more easily open it to look at it" before transmitting it to the Admiralty. Of his plans for 1775, Gauld wrote:

We are to set out in a few days for the Florida Kays and shall begin where we left off last year, continuing the Survey to the Eastward towards the Cape. It is very laborious and troublesome work among those Kays and Shoals, but as a thorough Knowledge of them is essential for the Navigation through the Gulf of Florida, the utmost care shall be taken to do them justice.[19]

About the first of May, Gauld began the year's work. "We began where we left off last year," he wrote, and "carried on the Survey as far as Cayo Largo, to what I believe may properly be called Cape Florida in the Latitude of about 25 degrees North. We met with no good Harbours this year for large vessels, but a great many dangers to which they may be exposed." *Florida* was at Key West by May 23, and she soon began working eastward past Sambo Key (modern Boca Chica). On June 30 she was at Cayos Vacas (Vaca Key), where she remained until mid-August. The base vessel then moved on to New or Upper Matacumbe for two weeks, and in mid-September Lieutenant Cobb anchored off Kay Tavernier while Gauld extended his investigation to Key Largo. The survey ship returned to Key West at the end of the month, and by October 14 the whole party was back at Pensacola. Gauld was delighted to have gotten such an early start and to have returned to Pensacola "in good Season before the cold stormy weather set in."

18. Gayton to Admiralty, January 5, 1775: ADM 1/240.
19. Gauld to Gayton, April 23, 1775: ADM 1/240. See also captain's log, *Diligence:* ADM 51/25; *Florida* muster table: ADM 36/7978.

Exploring the Limits and Beyond, 1774-1778

The unceasing routine of the surveying team had twice been broken by the unexpected appearances in the keys of the sloop H. M. S. *Savage* and the schooner *Saint John* bound from Providence to Pensacola. "Both of them went all the way between the Reef and the Kays," Gauld reported, "and as we happened to be at Cayo Huesso [Key West] when the *Savage* was passing by, Mr. Payne piloted her into the Harbour, where we then lay, at the desire of Captain [Hugh] Bromadge, who did not expect to find a Harbour there." The surveyor was obviously delighted at making the results of his labor immediately useful to one of His Majesty's captains. The attractions of Key West deserved closer attention: "it were greatly to be wished that it was more generally known on account of its vicinity to the Havanna, as Frigates and even fifty Gun Ships might easily go in and anchor there in great safety."[20]

The encounter with *Savage* must have been particularly pleasing to George Gauld, for Captain Bromedge carried a distinguished passenger, none other than former Lieutenant Governor Montfort Browne, who had left West Florida in disgrace in 1770. Since then, Browne had won exoneration and appointment to the governorship of the Bahamas — but he had not lost interest in West Florida. In the spring of 1775, using his health and the far-flung extent of his island government as an excuse, he made "a little excursion . . . by Sea" in H. M. S. *Savage* and extended it as far west as Pensacola. His visit doubtless gave him an opportunity to triumph over some old enemies — a pleasure Montfort Browne would have enjoyed to the fullest — but he was no less interested in the development of the colony's western lands. He had long had his eye on the Mississippi and at one time had offered himself as governor should a new province be carved out of West Florida; even the Spanish authorities at New Orleans knew of Browne's ambitions in that direction. For the moment

---

20. Gauld to Gayton, November 17, 1775: ADM 1/240, printed in William B. Clark, ed., *Naval Documents of the American Revolution* (Washington, D.C., 1966), 2:1062-63; muster table, *Florida:* ADM 36/7978; rough chart: MODHD, D959a 88. Gauld's account of meeting *Savage* and *St. John* is puzzling. Virtually all of *Savage*'s records have disappeared, but Gauld's encounter with her must have occurred toward the end of May, for she returned to New Providence from Pensacola on July 19, 1775. Lieutenant William Grant of *St. John* reported encountering a survey sloop from the Jamaica station off Tortugas Key, August 15, at which date *Florida* was at Cayos Vaccas. Robert R. Rea is unable to explain the apparent discrepancy; both Grant and Cobb certainly knew where they were, and *Florida* had left Key West on her homeward voyage before *St. John* could have gotten there on her return voyage to the Bahamas in October. Captain's log, *St. John:* ADM 51/4330.

his attention centered upon the prospective capital of the proposed Mississippi colony, a town to be named Dartmouth in honor of the Colonial Secretary and to be located (as Browne suggested) on property owned by Montfort Browne to the north of modern Vicksburg and known as Walnut Hills. To support his case, Governor Browne consulted with George Gauld (whom he erroneously described as "the Surveying Engineer of this district for their Lordships of the Treasury"), Thomas Hutchins, and others. The would-be realtor secured from Gauld a draft of that portion of the Mississippi River bank which he wished to sell to the home government; a copy of the pertinent section was made by Nathaniel Lindegren of the 16th Regiment and dispatched to Lord Dartmouth. Thomas Hutchins lent his name to Browne's scheme, but George Gauld remained silent.[21] Browne might boast that the Mississippi River at Vicksburg (or Browne's Cliffs) would shelter a thousand ships "from any winds, current, or Loggs," but Gauld's Key West was a present reality and no doubt a most welcome haven to the governor's sloop.

The accuracy of Gauld's drafts of the Florida keys, ultimately embodied in *An Accurate Chart of the Tortugas and Florida Kays*, may owe something to his friend John Lorimer. In 1790, when William Faden issued the chart and the accompanying *Account of The Surveys*, he remarked — evidently on the basis of Gauld's manuscript notes or from information provided by Dr. Lorimer himself — that "the *Dip* or inclination of the Magnetic Needle was . . . taken in the course of these surveys, by an instrument of peculiar construction." The publisher failed to elaborate upon this mysterious piece of equipment, but it seems likely that it was a "dipping needle" designed by Lorimer and constructed for him in England by a Mr. Sisson. With such an instrument Gauld might have established the line of magnetic variation shown running through Cabbage Tree Island and carrying the date 1775. Lorimer was well aware of the difficulty of using even the best dry-card magnetic compass aboard ship, and George Gauld would have welcomed anything that promised greater precision in establish-

---

21. Browne to Dartmouth, Pensacola, June 12, 1775, and enclosures: C.O. 23/23: 7369. Browne's letter implies that he consulted with Gauld at Pensacola. Considering the known chronology, the fact that Gauld's chart was certainly copied for Browne at Pensacola by Lindegren (a most unlikely procedure had Gauld himself been on hand), and Governor Browne's well-known tendency to stretch a point, it may be guessed that Gauld and Browne met at Key West, and there the surveyor gave the governor permission to have his recent rough draft of the Mississippi copied when Browne reached West Florida.

ing bearings and exact locations. The surveyor's "instrument of peculiar construction" was doubtless Dr. Lorimer's "Universal Magnetic Needle or Observation Compass," whose description appeared in the Royal Society's *Philosophical Transactions* in 1775. Unfortunately, Lorimer wrote the article in 1773, so it contains no reference to field use of the compass by Gauld, then working out of Port Royal; but the date on Gauld's chart and Faden's reference, taken together, strongly suggest that the surveyor took Dr. Lorimer's compass, its gimballed brass rings set in a handsome wooden chest, with him on his last voyage to the Florida keys.[22]

The surveying sloop *Florida* returned to Port Royal on January 15, 1776, for the customary annual refitting. The concerns of the British squadron based on Jamaica, like those of the whole British empire in North America, were already being affected by the revolutionary outburst to the north. Admiral Gayton ordered *Florida* "to be Careen'd as fast as possible" and planned then "to employ her in the best manner for His Majesty's Service 'till 'tis time for her to return to the Surveyor."[23] Just as there were minutemen behind the fences and hedges of Massachusetts, so were there daring Yankee skippers ready to fall upon British shipping in the West Indies, former smugglers now turned patriot privateers who knew the keys and coves better than the handful of Royal Navy captains on whom Gayton must depend for the protection of British trade and commerce; moreover, their shallow-draft vessels were better designed for inshore operations and the cutting out of helpless merchantmen as they sheltered in the island bays and took on wood and water. Their skill and daring was made embarrassingly evident in March 1776 when they descended upon the Bahamas and combined insult with the extensive injury of their pillaging by kidnapping unfortunate Governor Montfort Browne. Only two British ships escaped capture by the Americans: *St. John*, which Gauld had encountered the previous summer, and a sloop whose master, Francis LeMontais (Admiral Gayton's nephew), would eventually appear at Pensacola as commander of H. M. S. *Stork*. Not only were the islands and shipping lanes threatened; there was also the long boundary of rivers and lakes separating

22. *An Account of The Surveys*, p. 21; "Description of a new Dipping-needle. By Mr. J. Lorimer of Pensacola, in a Letter to Sir John Pringle, Bart. F.R.S. September 13, 1773," Royal Society, *Philosophical Transactions*, vol. 65, pt. 1 (1775): 79–84.

23. Gayton to Admiralty, January 21, 1776: ADM 1/240; muster table, *Florida*: ADM 36/7978.

G^eo. Gauld — Cartographer

British and Spanish America that cried for suitable vessels to patrol its waters and protect the interests of loyal settlers and merchants.

The Admiralty was aware of these problems and did what it could to meet them. On February 16, 1776, Gayton was authorized to purchase four small vessels to be employed on the lakes and the Mississippi River; one of these was to draw no more than six feet of water and was to be named *Florida*. By mid-June the Admiral had purchased three vessels including "a sloop of 60 tons and drawing Six feet Water abaft which I have call'd the *West Florida* (to make a Distinction between her & the Florida Surveying Sloop)." Gayton proposed to send her to the Mississippi, as directed, and placed her under the command of one of the junior officers aboard his flagship, Lieutenant George Burdon.[24]

The admiral could ill afford to leave even so modest a vessel as the surveying sloop unemployed. At one point Gayton found that there was no other ship at Port Royal in which to send home vital dispatches, save *Florida*: "which had I done it, Mr. Gauld could have done Nothing this Year in Surveying." With some acerbity he added, "And I suppose would have received his £365 from Government; the same as if he had been actually employed." The admiral's express was delayed, but Lieutenant Cobb spent some time in March cruising off Cape François, Hispaniola, and Gayton assured Their Lordships that in spite of the awkwardness of maintaining a civil operation in time of war, he would fit and send *Florida* to the surveyor "with all possible dispatch." *Florida's* peaceful employment posed a further problem for the C-in-C Jamaica, for her crew was still being drafted out of the other ships at Port Royal. Gayton was as short of men as of ships and urged the Admiralty that the surveying sloop should be put "on the same Establishment as the other Arm'd Vessels as it much weakens the few Ships I have, her being Manned out of them."[25]

When *Florida* reached Pensacola, probably about mid-May,[26] George Gauld had completed a fair copy of the 1775 survey of the keys and was ready to send it home by way of Port Royal. As for the current season, Gauld advised the admiral, "It is generally thought here, and indeed both Mr. Cobb and I are of the same opinion, that it

---

24. Gayton to Admiralty, June 13, 1776: ADM 1/240; in Clark, *Naval Documents*, 5: 521–22.

25. Ibid.; Gayton to Admiralty, March 28, 1776: ADM 1/240; in Clark, *Naval Documents*, 4: 553.

26. Cobb to Gayton, May 21, 1776; in Clark, *Naval Documents*, 5: 196.

would be very imprudent to proceed among the Florida Kays, where we left off last year, considering that New Providence has lately fallen into the hands of the Americans, and their Privateers are sufficiently acquainted with the Kays to annoy us if they think proper, as they will no doubt expect us there." It had been Gauld's hope to proceed from Cayo Largo northward to Cape Canaveral and thence to the Bahamas, but he was sure that "we can find employment enough for the present without going there."

A hard Gale of Wind that happened some years ago has greatly altered some parts of the Coast to the Westward of Pensacola, since I surveyed it, particularly Ship Island and the Chandeleurs. Wherefore we think it will be most proper in the present Situation of affairs, first to examine those places and the Mouths of the Mississippi; then to try how far the general Bank of Soundings runs off along the Coast, and afterwards we intend to carry on the Survey of the West part of the Peninsula of East Florida, near the Bay of Espíritu Santo, where we left off when I was ordered up to Jamaica by Sir George Rodney to survey the Harbours of Port Royal and Kingston.

Not for the first time Gauld pleaded for the materials necessary for his work:

I am greatly in want of large Drawing Paper, and black-lead Pencils, particularly the largest kind of Paper called Double Elephant. I wish their Lordships would be pleased to order out a Box of Stationary, as I have had but one since I was first engaged in this employment, with now and then some occasional supplies from Jamaica. At present there is even no common writing paper to be got here.

As was his custom, the surveyor forwarded to Port Royal certificates of service on which his salary would be paid in London. It was an awkward arrangement at best, and in 1776 Gauld begged Gayton to be particularly expeditious in forwarding his certificates to the Admiralty because "My late Agent has broke [i.e., taken bankruptcy] with almost all my money in his hands, which has distressed me very much, as some Bills to a considerable Amount have come back protested."[27]

To what extent George Gauld was able to carry out his plans for

---

27. Gauld to Gayton, May 20, 1776: ADM 1/240; in Clark, *Naval Documents*, 5: 176–77. *An Account of The Surveys*, p. 27.

G^{eo.} Gauld — Cartographer

1776 is uncertain; no account of his activity seems to have reached the Admiralty. *Florida* sailed from Pensacola during the first week of June and was off the mouth of the Pearl River during the latter half of the month. In July she was at Ship Island, and there, on the twentieth, John Payne was removed from Gauld's vessel to serve as a pilot aboard H. M. S. *Diligence*. *Florida* was in Nassau Road at the end of the month and returned to Pensacola in about two weeks. Payne's presence aboard the survey vessel suggests that Gauld was pursuing the first part of the plan he had outlined for Admiral Gayton, but he certainly did not return to the East Florida coast. Payne, on the other hand, was able to add something more to the survey team's efforts in 1776. *Diligence* was cruising among the keys in August, and Payne, provided with one of Gauld's earlier rough charts of the area, took a number of careful sightings for the purpose of adding to and correcting the naval surveyor's work. He was able to submit his findings to Gauld when he returned to Pensacola in September. The surveyor would sail again with John Payne, but this summer marked the end of an even older relationship. On September 17, Gauld's servant, Peter Quash, was discharged from the service at Pensacola. In his place Gauld took on the services of Charles Robinson, who continued with him for the next three years.[28]

Wild rumors of an American invasion of West Florida were circulating, and Governor Chester and his council were inclined to keep the protective shield of the Royal Navy close at hand. At the moment, Lieutenant Cobb's little *Florida* was the only naval vessel regularly stationed in the harbor, and as the senior commander present, Cobb inquired of Governor Chester, on September 2, what use he might suggest for his former midshipman George Burdon and *West Florida*. The council replied, two days later, proposing that the only fighting ship in provincial waters should be used to transport lumber from the naval establishments at the Red Cliffs (Barrancas), Santa Rosa Island, and Tartar Point to the town for the strengthening of the fort. They also recommended that Cobb should keep *Florida* "in readiness immediately to proceed to Jamaica . . . with dispatches" for the governor and commander in chief "representing the Alarming Situation we are at present in, and require all assistance which they

---

28. Muster tables of *Florida* and *Diligence:* ADM 36/7978 and 7586; captain's log, *Diligence:* ADM 51/25; notes on the rough chart of the Dry Tortugas and Keys: MODHD, U9 6g. No further identification of Gauld's new servant has been found, but the surveyor once referred to him as "Dr. Robinson."

can afford us may be immediately sent down for the Relief and protection of this Province."²⁹ Cobb acquiesced to the degree of holding *West Florida* in harbor until another naval vessel should arrive.

Lieutenant Burdon was more interested in provisioning his ship for sea duty than in serving as an army lighter, and at the end of the first week of September he worked his way down the bay toward Santa Rosa Island and the mouth of the harbor where any defensive naval action might be expected to be fought and where he could keep an eye on passing vessels. On September 15, the sloop *Diligence* entered the bay. Captain Thomas Davey's seniority enabled him to suggest to the governor that it was no longer necessary to hold Burdon in check; if the civil authorities would agree, he would send *West Florida* to the lakes according to Admiral Gayton's orders. The council was persuaded that the presence of *Diligence* justified the release of *West Florida*, and on October 20, Burdon crossed the bar and headed west. An express was indeed sent to Jamaica by the governor of West Florida, relating the rumors on which his fears were based, but the provincial schooner rather than H. M. S. *Florida* ran Peter Chester's errand. Lieutenant Cobb's surveying ship remained in harbor until the end of the year, when she returned to Jamaica for her annual refitting.³⁰

With the passage of time, the arrival of naval reinforcements, and the failure of an American attack to develop, the government of West Florida began to relax its fears. The revolution was far away; West Floridians were a loyal lot, by and large, and loyalism was strengthened — or so it appeared — by a great influx of refugees fleeing from the troubled colonies to the north. In 1777, *Florida* was not needed for the defense of the provincial capital, and George Gauld might resume his work. The sloop returned to Port Royal during the winter of 1776–77, and a new commander, Lieutenant John Osborn, was aboard when *Florida* returned to Pensacola on April 11, 1777; but Gauld would once more enjoy the assistance of John Payne when he sailed on what proved to be his last surveying expedition.

Adhering to an ideal schedule, the surveying party left Pensacola some time after the middle of April, and as in the past Gauld had at his disposal *Florida*, the old surveying boat, and a shallop. Quite unexpectedly, severe spring storms pounded the Gulf Coast during the

29. Council Minutes: C.O. 5/631.
30. Ibid., September 21, 1776; captain's log, *West Florida:* ADM 51/4390; captain's log, *Diligence:* ADM 51/25; muster table, *Florida:* ADM 36/7978.

last ten days of April. The shallop foundered; the surveying boat, which *Florida* took in tow, "very near shared the same fate, and the Sloop was likewise in a dangerous Situation."[31] Happily no lives were lost, but Gauld and Osborn were obliged to return to Pensacola before May 1 for repairs and were there May 3 when *West Florida* returned to the harbor in similarly battered condition.[32]

By May 13, the surveying ship had completed its repairs, and Gauld and his colleagues made a second start "with an intention to compleat everything to the Westward. I had heard by late reports," wrote the surveyor, "that the Southwest Pass of the Mississippe had now the deepest water, some said Seventeen feet. I was therefore determined to take this Opportunity of reexamining it, as I could not find Ten feet in it when I surveyed the Mouths of the Mississippe some years ago."[33] Current information regarding channel depths was of critical importance, for in order to intercept contraband intended for the colonies in rebellion, British men-of-war were now keeping a sharp watch on shipping moving in and out of New Orleans.

Florida apparently proceeded directly to the delta, and "having carryed the Sloop into the SW Pass and securely moored her to two Willow Bushes," Gauld left Lieutenant Osborn posted at the entrance while he pursued a more interesting investigation. During the three weeks *Florida* remained on station, H. M. S. *Atalanta* appeared, having gone up to New Orleans where she remained until May 12, when the rumor of an American privateer at the mouth of the river brought her downstream in search of prey. At the same time, Lieutenant Burdon in *West Florida* was cruising Lake Pontchartrain in successful pursuit of smugglers plying between British and Spanish territory on either side of that lake. The long arm of the Royal Navy seemed, in the summer of 1777, to encircle the throat of Spanish Louisiana, and the visible evidence of its strength encouraged George Gauld to extend his exploration of the Gulf Coast beyond the bounds of British sovereignty.

"The weather being very fine for that purpose," Gauld wrote, "Mr. Payne and I took a three Weeks Cruise in the Boat to examine

31. Gauld to Gayton, December 30, 1777: ADM 1/240.
32. The foregoing chronology is established by the captain's logs of *West Florida:* ADM 51/4390 and *Hound:* ADM 51/463; muster table, *Florida:* ADM 36/7978; Burdon to Gayton, May 14, 1777: ADM 1/240.
33. Gauld to Gayton, December 30, 1777: ADM 1/240.

Exploring the Limits and Beyond, 1774-1778

the Coast a little to the Westward of the Mississippe."[34] The surveyor's party sailed along the Louisiana coast at least sixteen leagues west of the pass where they left *Florida*, for at that distance Gauld discovered the only bay on the Louisiana coast that offered shelter even for small boats. His harbor, denoted by symbolic anchors on his chart, was present-day Terrebonne Bay, recognizable but much changed after the passage of two hundred years. Timbalier Island appears to have disintegrated considerably at both ends, leaving Timbalier Bay more exposed to the sea than it was in 1777 and allowing northerly access to Terrebonne Bay, where Gauld found the channel to run to the northwest. He located and identified Calumet Island but placed it considerably to the westward, in Terrebonne rather than Timbalier Bay.[35]

When their supplies were exhausted, perhaps about June 10, Gauld and Payne returned to the sloop's anchorage at the mouth of the Mississippi and "took a very particular Survey of the SW Pass where we found at least Fourteen feet of water," a change in depth which was duly noted on Gauld's chart. Now he must make a major decision, one to which his whole career pointed. It could not have been difficult. As Gauld put it to Admiral Gayton:

We then proceeded in the Sloop to the Westward to take a further Sketch of that unknown and unfrequented Coast. For although we had no right or authority to survey beyond the Mississippe, yet a general knowledge of that Coast is absolutely Necessary for Vessels bound to the River, & as we could not with safety proceed amongst the Florida Kays on account of the Rebel Privateers, we just snatched this Opportunity, as I was very anxious to have the Sea Coast at least as far to the Westward, as the furthest part of the River where I had been [northward], *on purpose to make the Plan Square.*[36]

The cartographer's expressed desire to balance the dimensions of his map displays a certain artistic sense, no doubt, but his recognition of the centrality of the Mississippi and New Orleans suggests a clear grasp of its strategic importance to the whole Gulf Coast re-

---

34. Ibid.
35. "A Draught of Part of the Coast to the Westward of the River Mississippi with Part of the Island of New Orleans &c. By George Gauld M.A. For the Right Honourable The Board of Admiralty 1777." MODHD, D965 88; cited hereafter as "Draught 1777." Comparison has been made with C. & G.S. Chart 1116, Mississippi River to Galveston.
36. Gauld to Gayton, December 30, 1777: ADM 1/240.

gion. At the end of the last great war for empire, in 1763, the peacemakers had lacked adequate maps on which to trace the boundaries they were drawing; should history repeat itself in the next Anglo-Spanish war (an event not to be doubted in the Royal Navy), George Gauld would see to it that proper charts were ready and at hand. By July 8, *Florida* was sailing westward, exploring the coast of Spanish Louisiana.

"In the Course of our Cruise in the Sloop," Gauld wrote, "we tryed every opening or Inlet as near as we could approach but found no Harbour, except one for small Vessels wch. Mr. Payne & I discovered in our first Cruise in the Boat." On past Timbalier and Terrebonne bays they sailed, on past the Isles Derniers (more solidly impressive than today) to Raccoon Point, as it still appears on charts. Across these islands toward the north, the surveyor could only record "Innumerable Islands and Lagoons as far as the Eye can reach from the Top-mast Head." At Raccoon Point, July 11, Gauld took the latitude at 29°02' N, scarce a minute off that provided by modern charts. Further west he noted the prevalence of the mollusks that gave their name to his Oyster Point and modern Oyster Bayou. He sketched but gave no name to Point au Fer and noted the shell reef as "Oyster Banks mostly dry at Low Water."

Proceeding into Atchafalaya Bay, Gauld worked up the river half a dozen miles and remarked, "This River, from the Colour of the Water, seems to be a Branch of the Mississippi, perhaps the Chafalaya or Apelousa. It is broad but has very little Current. The Water is deep inside, but there is such a Ledge of Oyster Banks between it and the Sea, that it is hardly possible to find a Channel even for a small Vessel into it." Then or later he worked back into Vermillion Bay and determined the outline of Marsh Island, noting the offshore Shell Keys and Tiger Shoal. Further west Gauld noted the mouth of the Mermentau River, passed an old wreck, and came to the mouth of the "Catcatchouk" or Calcasieu River on July 20. Here he identified "the Northernmost Part of the Gulf of Mexico, to the Westward of the Mississippi," although his latitudinal measurement was in error by some five minutes. He hinted rather accurately at the outlines of Calcasieu Lake and the river beyond which leads to modern Lake Charles, Louisiana, and he reported that "There are several Indian Settlements on this River."[37]

37. Gauld, "Draught 1777."

Exploring the Limits and Beyond, 1774–1778

On they pressed, some eighty leagues west of the Mississippi, and were about to reverse their course when they chanced upon the scene of one of those marine tragedies that more than justified George Gauld's explorations. Just within the mouth of the Sabine River, near modern Louisiana Point, the surveyor recounted:

We found the Wreck of a Sloop, near the Entrance of a River called by the Indians Chicoanche, & took on board three men belonging to her, the Captain, Passengers, & all the rest of the People being Dead. The Savages had stripped the Vessel of her Sails & everything they could carry away. They had sailed in Nov. 1776 from Montago Bay [Jamaica] bound to the Mississippi, but falling in to the Westward they had bewildered themselves on that desolate Coast, & were cast away. The Sloop was called the Robart, belonging to Messrs. Thompson & Campbell.[38]

Gauld subsequently changed the spelling of the river to Chicouansh, on his chart, sketched and sounded Sabine Lake, and indicated the river that flows south from Orange, Texas. He added a note on the chart relating that there had been nine men aboard the ill-fated sloop *Robert*. Of the survivors, he wrote:

These three Men in a small Boat wandered along the Coast for some Months in quest of the Mississippi; but after a fruitless search they had returned to the Wreck for some Provisions, and were just going away again when providentially the Surveying Sloop Florida appeared and relieved them from their distress July 22nd 1777, after they had been eight Months from Jamaica.[39]

At the mouth of the Sabine River, Gauld decided to divide his forces:

I took a Months Provisions in the Boat to return & examine the Coast more particularly as far as the small Harbour . . . where Mr. Payne & I had been before, as the Sloop could not go within two or three Leagues of the Shore in many places . . . Mr. Osborne & Mr. Payne were to proceed in the Sloop a little further to the Westward to make what discoveries the time would permit, & meet me at the little Harbour. Mr. Osborne who has always shewn the greatest readiness to forward this Service willingly concurred in this Plan of Operations.

38. Gauld to Gayton, December 30, 1777: ADM 1/240.
39. Gauld, "Draught 1777."

G^eo. Gauld — Cartographer

The last week in July, then, *Florida* sailed westward along the Texas coast, covering "about 18 or 20 Leags. & found the Shore trenching much to the Southward, & Mr. Payne continued to Sketch of the Coast as far as they went."[40] Osborn took his ship along the "Low Sandy Beach" of Bolivar Peninsula, past the entrance to Galveston Bay, whose southwestern tip he christened Indian Point because of an "Indian flag" observed there. A few miles farther (at 29°11′ on Gauld's chart), on August 1, they turned back to make their rendezvous with Gauld. The naval surveyor duly noted on his chart that "Mr. John Payne laid down the Chicouansh and the Coast to the Westward of it." Payne's line of soundings along the coast ends at the mouth of Galveston Bay. He looked north-northwest into that body of water and reported "No Land to be seen in this Direction from the Mast head." *Florida*'s return voyage took her roughly parallel to the Texas-Louisiana coast, sounding as she went in order to provide warning of depths and bottom conditions to mariners approaching that "inhospitable Coast."[41]

Toward the end of August, Osborn and Payne reached their rendezvous at Isle au Calumet. There was no sign of George Gauld, and there must have been a few nervous days before the surveyor's boat slipped into Terrebonne Bay. Directed by Gauld once more, the expedition "Return'd to the SW Pass again intending to run up the river to buy some Stock & refreshments, as we could not spare time before. This was the beginning of Sept. In our way up the River we had like to have had a Scuffle with a Spanish Packet for their Insolence." Unfortunately, Gauld depended upon Lieutenant Osborn to inform the commander in chief at Jamaica of this affair and provided no details — and Osborn's report seems to have disappeared. A month later, probably after he returned to Pensacola, Gauld learned something of the repercussions of the episode:

In consequence of a misrepresentation of the matter by the Capt. of a French Brig (who was likewise very abusive as he past by us just as we was Bearing down on the Spaniard) the Governor of New Orleans (who, in several of his Actions toward the English on the Mississippi, seems to have been rash & ill advised) took upon him to send down an Armed Brig

---

40. Gauld to Gayton, December 30, 1777: ADM 1/240.
41. Gauld, "Draught 1777." Exact locations and dates are derived from the rough chart: MODHD, E. 19/14 4a.

Exploring the Limits and Beyond, 1774–1778

with a large Party of Soldiers on purpose to take or destroy us. Perhaps it was lucky for both them & us that we were gone before they came, for Mr. Osborn never would have suffered with impunity any Indignity to be offered to the Kings Sloop.[42]

Thus George Gauld first encountered the new Spanish governor of Louisiana, thirty-year-old Bernardo de Gálvez, who had assumed command at New Orleans at the beginning of the year. There had indeed been "rash & ill advised" clashes between young Gálvez, and the British while Gauld was peacefully pursuing his trade in Spanish territorial waters. In response to *West Florida*'s activity on Lake Pontchartrain, Gálvez had seized the boats of British merchants doing business at New Orleans, ordered them out of his territory, and rejected the protests delivered in May by Captain Thomas Lloyd of *Atalanta* and by Major Alexander Dickson in August. Gálvez also cooperated with Oliver Pollock so that American officers were able to ship guns and powder up the Mississippi and Ohio rivers to Fort Pitt.[43] British West Florida would learn more of Governor Gálvez, to its regret, but as yet neither antagonist was desirous of open hostilities.

"Having gotten what we wanted in the River," Gauld continued, "we made the best of our way to the Isle au Breton, from whence the Sloop went to Pensacola for a Supply of Provisions [September 15–22].... We took 5 weeks Provisions in the Boat, that I might have perhaps the only opportunity of compleating several parts about the Skirts of the Island of New Orleans wch. on any other Occasion would have hardly been deemed worth the time & trouble."[44] Lieutenant Osborn made a quick trip to Pensacola and sailed from there about September 25 to rendezvous with Gauld at Ship Island; he arrived on October 1, ahead of the surveyor.[45] The "skirts" of Spanish Louisiana that Gauld wished to tidy up included the shores of Lake Borgne, and there, on September 18, he "fell in with Mr. Burdon" and *West Florida*. "It was very lucky for him that we happen'd to come that way," Gauld reported, "for he could hardly have got to Pensacola, his Sloop being very Leaky."[46] Gauld was not exag-

42. Gauld to Gayton, December 30, 1777: ADM 1/240.
43. John Walton Caughey, *Bernado de Gálvez in Louisiana 1776–1783* (reprint, Gretna, La., 1972), pp. 67–76, 88–92.
44. Gauld to Gayton, December 30, 1777: ADM 1/240.
45. Osborn to Gayton, January 17, 1778: ADM 1/240.
46. Gauld to Gayton, December 30, 1777: ADM 1/240.

gerating. Lieutenant Burdon's crew had been depleted when he seized and manned a prize in June; he lost five deserters in July and August; his ship was making more than four inches of water every hour, and "the fevour Augue [was] so bad on board that only the Lieutenant, Mate, and one Man [were] well" at the beginning of September.[47] The surveying party "lent a hand" in the best sense of the term, and both Burdon and Gauld worked their separate ways back to Ship Island, where *Florida* was found waiting on October 4. Gauld returned aboard his sloop on October 8, by which time the two navy vessels and Burdon's prize had been joined by the merchant brig *Two Brothers*, and the little flotilla sailed east for Pensacola. George Gauld was home by October 16.[48]

The surveyor would normally have submitted his report of the year's work soon after returning to Pensacola, but the ship by which he intended to send his letter to Admiral Gayton sailed without warning. It was not until January 19, 1778, that he was able to seal his papers, declaring to the commander in chief, "I have now done everything that was wanted to the Westward and shall make out a fair plan to be transmitted to the Board of Admiralty." He did not propose to take any chances with the new chart, however, and announced that he would only send it by a man-of-war sailing either to Jamaica or directly to England.[49]

If George Gauld was pleased with his accomplishments at the end of the 1777 surveying season, Lieutenant John Osborn was not. *Florida*'s complement had evidently been increased for the 1777 survey, and he had found the sloop "incapable of carrying a Sufficient Quantity of Provisons, which Occasions the People being at Short allowance during the Summer, when they undergo the greatest fatigue; and as the Service . . . requires her to carry a great Quantity of Stores, her hold is so full that the People are oblidged to sleep on Deck." Inconvenience and discomfort were the least of Osborn's problems, however. Throughout the summer, he reported to Admiral Gayton, "the Sloop leaked much, & complained in every part, but

47. Captain's log, *West Florida:* ADM 51/4390.
48. Ibid.; Gauld to Gayton, December 30, 1777; Osborn to Gayton, January 17, 1778: ADM 1/240; muster table, *Florida:* ADM 36/7978.
49. Gauld to Gayton, December 30, 1777: ADM 1/240. In spite of Gauld's caution, the chart resulting from his work in 1777 was very nearly lost, for it was aboard H. M. S. *Hound* when the hurricane of October 9, 1778, struck Pensacola. Gauld to Parker, January 14, 1779: ADM 1/241.

Exploring the Limits and Beyond, 1774-1778

during this last passage [from Ship Island to Pensacola] we could hardly keep her free [of water] with the Pumps." Captain Thomas Lloyd of *Atalanta* was the senior captain at Pensacola when *Florida* returned in October, and upon Osborn's request he ordered the surveying sloop careened. She was hove down at Deer Point on December 15 and disclosed an appalling sight. Her entire sheathing was either destroyed by worms or simply ripped off; her bottom was eaten through by rats. Half her bow timbers were so decayed and rotten that "they might be pulled to pieces with one's fingers."[50] Captain Lloyd ordered her surveyed and received a report on January 1, 1778, that *Florida* was indeed "unfit to proceed to Sea, till she has a thorough repair, which cannot be given her at this Port."[51] Without further ado, the sloop was condemned. By March she had been sold to the army ordnance people to be used as a floating magazine, a storehouse for Indian presents, and on at least one occasion as a barge to haul timbers across the bay.[52]

Lieutenant John Osborn assumed that the old *Florida* would be replaced by a new surveying ship, and he urged Admiral Gayton to consider "that a Sloop [square-rigged] is a very improper Vessel for the Surveying service, as Accidents continually happen to the Boats in hoisting them in and out. In a Schooner or Brig [rigged fore-and-aft], those operations are performed with more ease & Safety."[53]

The commander in chief on the Jamaica Station valued George Gauld's work and dealt promptly with the problem. Gayton purchased another ship, a schooner as Osborn had suggested, and a shallop and instructed the officers and crew of *Florida* to return to Jamaica aboard H.M.S. *Hound*. Unfortunately, Gayton's orders reached Pensacola only six days before March 22, 1778, when *Hound* sailed in great haste on a special mission to the Mississippi, and Osborn found himself stuck in Pensacola.[54] Governor Peter Chester proposed that Osborn and his men should be sent to strengthen the complement of *West Florida*, which was patrolling the coast between

---

50. Osborn to Gayton, January 17, 1778: ADM 1/240.

51. Report submitted to Capt. Thomas Lloyd, *Atalanta*, January 18, 1778: ADM 1/240. A second survey was made by the warrant officers of *Hound* and *Sylph* on May 13, 1778: master's log, *Sylph*: ADM 52/2025.

52. Captain Joseph Nunn to Parker, March 21, December 4, 1778; Lloyd to Parker, March 26, 1778: ADM 1/241; captain's log, *Hound*: ADM 51/463.

53. Osborn to Gayton, January 17, 1778: ADM 1/240.

54. Gayton to Admiralty, April 20, 1778: ADM 1/240; Osborn to Parker, March 25, 1778: ADM 1/241; captain's log, *Hound*: ADM 51/463.

Mobile and Lake Pontchartrain; but the senior naval officer at Pensacola, Captain Joseph Nunn, refused to issue such orders because only five of Osborn's men were fit for duty, and two of them were Americans who might be expected to desert at the first opportunity.[55]

To Gayton's successor, Sir Peter Parker, fell the task of dispatching the new surveying ship to the Gulf Coast. By April 19 the schooner was ready, and as it replaced the condemned *Florida*, it too was christened *Florida*. Under the command of Lieutenant James Kirkland, the surveying schooner with its thirty-man crew sailed for Pensacola on May 5, 1778, in company with a transport carrying reinforcements for the West Florida garrison. Kirkland enjoyed an exceptionally fast passage and entered Pensacola Bay on May 27, proudly escorting not only the troopship but a prize he had taken just off the harbor. Two days later *Florida* sailed for Mobile "to cruise for a Party of Rebels" who were said to have "taken a Sloop & fitted her for a Privateer."[56]

George Gauld would get no service from the new surveying schooner. During July and August 1778, Lieutenant Kirkland cruised between Ship Island and the Rigolets, where he frequently crossed courses with Captain Nunn in *Hound*. By September 11, both ships were back in Pensacola. *Florida* was careened at Deer Point when Nunn received a frantic message from Kirkland asking him to send *Hound*'s space pump and twenty men to save *Florida* from sinking. Their efforts were unsuccessful. The schooner filled and settled in the shallow water off the careening wharf even while they were trying to patch her up. Captain Nunn condemned her as she lay and ordered her crew aboard his own ship.[57] So ended the brief career of the second *Florida*. There is no evidence that George Gauld ever set foot aboard the schooner that was his last surveying vessel.

The commander of the old *Florida* sloop, John Osborn, was in no position to smile at his successor's discomfiture. With *Hound* off to New Orleans in March, it was decided that Osborn and his crew should join H. M. S. *Sylph*. Osborn cooled his heels at Pensacola until *Sylph* anchored below the town on May 10. Then, in short order, a final survey was held on his old ship, and he was appointed to com-

55. Council Minutes, March 18, 1778: C.O. 5/631.
56. Parker to Admiralty, April 19, June 21, September 23, 1778: ADM 1/241; master's log, *Sylph*: ADM 52/2025; Council Minutes, May 28, 29, 1778: C.O. 5/635.
57. Captain's log, *Hound*: ADM 51/463; Nunn to Parker, December 4, 1778; Parker to Admiralty, January 12, 1779: ADM 1/241.

Exploring the Limits and Beyond, 1774-1778

mand the little sloop *Catherine,* which was assigned to transport troops and supplies to Fort Bute at Manchac in company with *Sylph* under Captain John Fergusson. The two ships sailed on June 1, and after being held up for six days at the Balise because of light breezes and strong countercurrents, they entered the Mississippi on June 11. Slowly they worked their way up the river and were within a few miles of New Orleans when, at three o'clock on June 20, *Catherine* struck hard on a submerged tree stump. In spite of all that pumps and buckets could do, she sank and fell on her beam ends an hour later. No lives were lost, and as the sloop had been secured to the bank before she foundered and the river was falling steadily at the time, strenuous efforts were made to recover not only the stores and guns she carried but the ship itself. On the twenty-seventh and twenty-eighth, Osborn and Fergusson attempted to heave *Catherine* upright, and failing that, to haul her into deeper water so that the Spaniards might recover nothing from her when the river fell — but without success. Already reduced to two-thirds rations, Fergusson gave up the salvage attempt on July 5 and began sailing and warping his way up the Mississippi. Osborn and his people remained aboard *Sylph* until August 9, when, well below Manchac, they put off in a bateau in order to see to the final delivery of the remains of their cargo. *Sylph* could get no closer than five miles to her destination; her crew were sickly, provisions short, wind and current set against her, and the Mississippi had fallen twenty-four feet, making ship navigation most dangerous. On September 9 Fergusson swung about and began to drop down the river, nearly losing his own ship on a submerged snag. All hands were back in Pensacola by September 26.[58]

The surveying vessels were gone, their captains without commands and their crews distributed between *Hound* and *Sylph.* Both Osborn and Kirkland had returned to Jamaica by December 26, 1778. Admiral Parker understood that a surveying vessel was "much wanted" and promised to "endeavour to procure a proper one as soon as possible and send her to Pensacola," but the slender resources and far-flung responsibilities of the Jamaica Station in time of war with France, and then Spain, defeated his intentions.[59] On May 18, 1778, George Gauld, surveyor, and his servant, Charles Robinson, were

58. Osborn to Parker, March 25, 1778: ADM 1/241; master's log, *Sylph:* ADM 52/2025; captain's log, *Sylph:* ADM 51/918. Gauld named the last stretch of river below Manchac "Sylph Reach" after this episode: *An Accurate Chart* (1803).

59. Parker to Admiralty, January 12, 1779: ADM 1/241.

discharged from the muster of the armed schooner *Florida*, which lay in the waters of Pensacola Bay, to the supernumerary list of the sloop *Sylph*.[60] It was a purely formal assignment, for John Fergusson's ship had other business to attend to than surveying. In May 1779, *Sylph* sailed from Pensacola with a convoy for England, and on the first of the month Gauld and Robinson were transferred from her to *West Florida*.[61] In one way it was a happy juncture, for *West Florida* had a new captain. On January 10, 1779, Lieutenant George Burdon left the ship, and the next day his fellow officers at Pensacola accepted the findings of a surgeon that he was suffering from "a long continuance of pulmonic complaints attended with a flux" and was incapable of command.[62] Poor Burdon was immediately replaced by Gauld's old comrade and assistant, John Payne, and on the first of February he took *West Florida* to sea to resume her patrol duties in the inland waterways of the province.[63] His friend's success in securing a ship of his own must have pleased George Gauld, and it would have delighted him to think, in May 1779, that some day he and John Payne would resume their long-postponed surveying of the coast they had come to know so well. Alas! Even then the fates were conspiring against them. George Gauld might draw a few more maps, but the West Florida coastal survey was at an end.

60. Muster tables, *Florida* and *Sylph*: ADM 36/7978 and 7843.
61. Muster tables, *Sylph* and *West Florida*: ADM 36/7843 and 9895.
62. Parker to Admiralty, April 2, 1779: ADM 1/242.
63. Master's log, *Sylph*: ADM 52/2025; captain's log, *Sylph*: ADM 51/918.

CHAPTER XI

The Surveyor at Home

The surveyor's annual progresses along the coast of the Gulf of Mexico represent his primary accomplishment as they were his official employment. For fourteen years he sailed those waters in sloop, schooner, and yawl, charting the islands, surveying the bays, and returning to Pensacola to compile and perfect the finished charts, which were duly forwarded to the Admiralty. From the ships' logs and from his own letters something of the routine of surveying can be recreated: the discomforts of heat, sweat, and labor when both food and water (if not rum) were in short supply; the dangers of shoals, snags, and hurricanes; and the frustrations of crowded, leaky ships pressed to their limits and beyond by men whose grim determination and suffering can only be imagined. The cost to government in terms of ships is easier to recount than the toll in men or in the life of any individual.

When George Gauld returned each fall to the quiet waters of Pensacola Bay and settled into the security of his own home, there was work to be done. One cannot study the great charts and maps without recognizing that their creation required excruciatingly painstaking application no less remarkable than the physical endurance involved in gathering the observations on which they were based. The purely technical triumph of plotting the coast from the mouths of the Mississippi to Galveston Bay on a chart some twelve feet long is awesome. The skill, the infinite care, and the consistently precise accuracy display more than craftsmanship; they reflect the spirit of a creative artist. Occasionally, indeed, the artist escaped the cartographer's restraints. Panoramic views of the Florida keys were drawn with a mariner's interests in mind, but those of Port Royal seem to seek release from the technical requirements of the charts they accompany, to show that the eye of the surveyor rose above his crude flags, smokes, and seamarks to catch a finer vision of the scene before him. There is also the handsome print entitled *A View of Pensacola in West Florida* that displays the harbor with its numerous naval and merchant vessels, the stockade fort, and the houses of several leading Pensacolans. Dedicated to Admiral Sir William Burnaby, the engraving bears Gauld's name and was "sold by T. Jeffreys in the Strand, London." Although it carries no date, this portrayal of British Pensacola must have been drawn about 1765, when Gauld had reason to flatter Admiral Burnaby as "Commander in Chief of His Majesty's Ships at Jamaica and in the Gulf of Mexico" and when the names of James Noble and James Macpherson were still prominent in provincial affairs.[1]

The surveyor's labors with pen and parallels were accomplished under conditions hardly less trying than those of actual surveying. We know nothing specific about George Gauld's house, which was undoubtedly also his workshop, but as he was a simple seafaring bachelor it would have been much like the rest of Pensacola's private dwellings, built in haste according to the local fashion by unskilled workmen using the most readily available materials. Cedar posts were usually erected to frame a house; if not closed with rough plank siding, the interstices were filled with tabby, a cement-like substance made up from crushed shells and sand. The roof would be thatched and far from waterproof. The floor of beaten earth might be covered but with wooden planks at best. Only the better houses boasted fireplaces and chimneys. Windows were at a premium, but the mapmaker would have considered his need for light and demanded more than the ordinary householder. Furniture was minimal, being very expensive, but again Gauld would require a great long table (no more than a few planks set across supports, of course) on which to draft his charts. If he worked at night, candles would provide but a dim light. They were always costly, and as ordinary tapers quickly lost their shape in Pensacola's summer heat, myrtle candles were preferred. Perhaps for Gauld oil lamps served better. In any case, multitudinous insects would distract him in warmer weather, and the pen-

---

1. The copy at the Historic Pensacola Preservation Board Museum bears a penciled date of 177–, but the dedication to Burnaby (who visited Pensacola in 1765 and left Jamaica in 1767), the reference to Macpherson (who left Pensacola about June 1765), and publication by Jefferys (who died in 1771) all point to the earlier date Robert R. Rea has suggested. Gauld may have been inspired to try his hand at a popular art form, and to submit it to Jefferys, by Elias Durnford. Between August 1764 and March 14, 1765, the London publisher issued a set of six views of Havana sketched by Durnford when he was a young engineer on the expedition of 1762. Like Durnford's, Gauld's sketch carries titles and explanatory material in both English and French. This connection might also explain Jefferys's possession and unacknowledged use of Gauld's 1765 survey of Espiritu Santo Bay in the 1769 edition of Stork's *Account of East-Florida* (*vide* chap. 13). Durnford's views are in the British Museum Map Room: K, CXXIII/28 a-f.

etrating cold of winter would discourage late hours at the drawing board because faggots for the hearth had to be cut on the eastern side of the bay and were exorbitantly expensive when transported to town.

Supplies and equipment were always in short supply, as indicated by Gauld's frequent pleas for paper suitable for drafting purposes. His employer, the Admiralty, did not provide for his needs, and there were no stationers in Pensacola. The surveyor was not alone in complaining that he lacked the basic requirements for correspondence, not to mention the practice of his trade; even the general commanding the Southern Department, Frederick Haldimand, ran short of proper paper on which to submit reports to headquarters.

In spite of trying circumstances and shortages, Gauld worked on, survived the summer surveys, produced his annual charts, and lived — no small achievement in colonial West Florida — to become a leading figure in the community. His public stature was shown by his repeated election to the General Assembly, and it may be assumed that when he disappeared from the roster of the Commons House it was at his own insistence that the work of the Florida coastal survey must come first. Gauld clearly enjoyed the favor and attention of each of the governors of West Florida, George Johnstone, John Eliot, and Peter Chester. He recognized the glaring faults of Montfort Browne, but the lieutenant governor appreciated the surveyor's professional virtues, as did Lieutenant Governor Elias Durnford. Johnstone and Eliot were both naval officers and the others military men; all freely appointed a number of army officers to the provincial council, and it is a little surprising that George Gauld, as the most permanent member of the naval establishment residing in Pensacola, was not appointed to that board. His standing in society and his contributions to the colony qualified him quite as well as most who were so elevated. It may be conjectured that Gauld's failure to achieve the highest place of honor was due to his devotion to his surveying duties, the lengthy periods of absence which they imposed, and his own moderate interest in the political scene.

He did not escape other public service that could better fit within the schedule of his work. In 1769, Governor Eliot appointed him justice of the peace of the town and district of Pensacola, and in 1777 and 1778 Governor Chester again listed him among the quorum. The duties of a justice were not onerous, but they could involve him in much of the minor legal squabbling of everyday life; they imposed

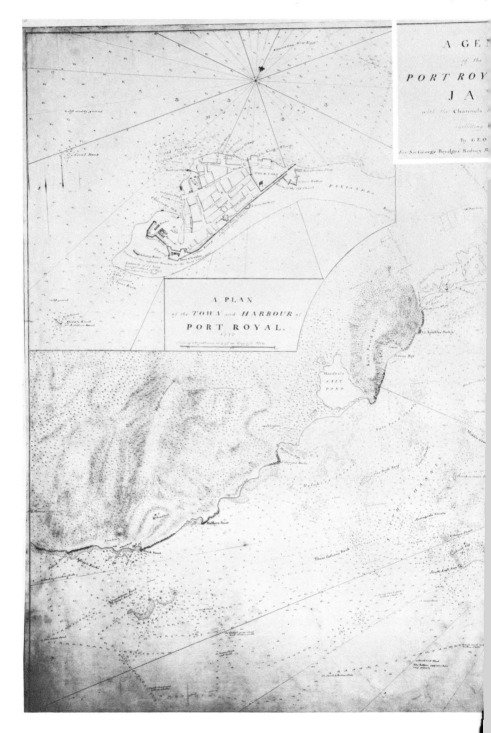

XIII. "A General Plan of the Harbours of Port Royal and Kingston." This magnificent chart reflects Gauld's work in Jamaica in 1772. Prepared for Admiral Rodney (and locating "Sir Geo. B. Rodney's Park"), it includes detailed plans of the naval

Neptune, in 1780. The richness of topographic detail contrasts markedly with Gauld's earlier manuscript charts. (Courtesy of the P. K. Yonge Library of Florida History.)

XII. *A Chart of The Bay and Harbour of Pensacola.* The only example of Gauld's work published during his lifetime, this chart appeared in Des Barres's collection, *The Atlantic*

base at Port Royal and the harbor defenses. (Courtesy of the Ministry of Defence Hydrographic Department.)

some obligations of availability to the public, and by the terms of colonial legislation they placed various responsibilities upon the individual. Justices were directed by their commissions "to keep and cause to be kept all ordinances and statutes for the good of our peace . . . and to chastize and punish all persons offending against the form of those ordinances and statutes." They were authorized to hear charges against those brought before them who might be accused of threats, assaults, and arson and to require that they post sufficient bond or be imprisoned. Justices were directed to inquire in their respective districts of all "manner of felonies, trespass, and of all and singular other misdeeds and offenses whatsoever committed by Negroes, Mulattos or Mustees," and their commissions empowered them to take statements under oath involving "felonies, trespasses, forestallings, regratings, ingrossings, extortions . . . and of all and singular other crimes and offenses." Their authority extended to "victuallers" caught in fraudulent practices involving the sale of food or abuses regarding weights and measures. Marshals, bailiffs, constables, and jailers were subject to correction by the justices for malfeasance and misfeasance in office. Finally, JPs were empowered to inspect indictments, make and continue processes, hold court, render judgment, and pass sentences involving virtually any crime except a capital offense; these they were required to certify to the chief justice who tried such cases in the General Court of Pleas.[2] As a considerable number of justices of the peace were appointed for Pensacola, George Gauld probably escaped much of the trivial but necessary burden the office imposed, but other duties he could not so readily avoid.

The General Assembly dealt with a variety of local problems, often inspired, no doubt, by individual members' experiences in local government as justices of the peace. In addition to assigning general duties of enforcement to the justices, the assembly also specified particular functions for boards named in its acts. Typically, the legislature in 1771 passed an appropriations bill which established a commission for providing and maintaining a pilot boat at Pensacola and named George Gauld among the six commissioners. The problem was one of great interest in a community that was dependent upon maritime trade and commerce not only for its prosperity but for its very life.

2. Council minutes, April 27, 1769, October 3, 1777, November 20, 1778: C.O. 5/631 and 635; Signs-manual, patents, commissions, Records of the States of the United States, Florida, West, E, Unit 3, E. lc, pp. 257–63, 307–11.

The Surveyor at Home

Charting a safe route into the harbor had been Gauld's first project, and no one had done more to provide mariners with a sure approach and specific knowledge of the bar and bay. In 1765, the merchants of Pensacola sought action from Governor Johnstone, and in 1768 they petitioned Lieutenant Governor Browne to establish a pilot at Mobile in order to encourage merchant vessels to utilize the ports of British West Florida rather than proceed on to New Orleans. Browne had agreed to appoint Captain Richard Hartley to that service and cited his qualification as having "attended Mr. Gall at his surveying a good part of the coast." In 1767, the Commons passed An Act Constituting Commissioners for the Examination and Appointment of Pilots, and for Establishing the Rates of Pilotage for the Harbour of Pensacola. Dr. John Lorimer shepherded the bill through the lower house, and in all likelihood he enjoyed the advice of his friend the naval surveyor regarding the pertinent details of the bill. Although it received its third reading in the Upper House, it failed to receive the lieutenant governor's approval at the end of the session because Browne found the funding provisions of the money bill unacceptable. Peter Chester was able to agree to a tax measure in 1771 which provided up to £25 "for providing a proper provincial pilot boat for Pensacola," so the designated commission was able to pursue its responsibilities. It is significant that George Gauld was the only member of the commission who was not currently a member of either the council or the Commons.[3]

Along with public recognition and service went the opportunity to gain wealth (at least on paper) through the easy acquistion of cheap land. The prospects were rather more illusory than real, but every West Floridian who possessed the means (and many who did not) speculated in colonial real estate. In addition to his town and garden lots in Pensacola, granted in 1765, George Gauld acquired property in modern Gulf Breeze, which was attractive to the eye of a seaman but possessed little commercial value two hundred years ago. Four years later, when he was a member of the General Assembly, Gauld was able to take advantage of a court-ordered sale, and through the good offices of Deputy Provost Marshal John Crozer he secured half

---

3. Council minutes, January 7, 1765: C.O. 5/625 and 632; Browne to Hillsborough, July 6, 1768: C.O. 5/577; Robert R. Rea and Milo B. Howard, *The Minutes, Journals, and Acts of the General Assembly of British West Florida*, pp. 81–101, 394.

of Pensacola town lot number 109, a tract with forty feet of frontage on Bute Street, located on the east side of town. On June 12, 1769, Gauld acquired temporary rights to the property for the consideration of a mere five shillings and an annual rental of one peppercorn; on January 5, 1770, the transaction was completed by a cash payment of 151 milled or Spanish dollars.[4] In 1772, while he was in Jamaica, Gauld received the thousand-acre grant on Thompson's Creek which he and Lorimer visited in 1774. This property was no doubt interesting but chiefly from a speculator's point of view, for the Mississippi lands were too far from Pensacola to be managed by an absentee landlord, and his naval duties bound Gauld to the Gulf Coast. These same obligations also gave George Gauld a uniquely thorough knowledge of the coastal lands of West Florida and the ability to identify properties that combined both agricultural potential and easy access by sea. In the years after 1770, the coast of the future state of Mississippi came under the scrutiny of serious investors, and it was there Gauld added to his holdings. He knew the offshore islands of Mississippi Sound, the river estuaries, and the settlements on their lower reaches; he was able, as few others could, to make a wise and enlightened judgment of the lands adjacent to these waterways. On May 24, 1776, Gauld submitted to Governor Chester and his council a document stating:

that your petitioner having resided chiefly in this province since its first establishment and being desirous of making a settlement on the east branch of the Pearl River . . . requests that two thousand acres of land may be granted him. Your petitioner begs leave humbly to mention not only his service for the good of the province, as surveyor of the harbours and seacoasts, but likewise several other extra services for which he never had nor required any consideration, though executed at considerable expense and trouble to himself. And as your petitioner has formerly made application for very little land in the province, he therefore humbly hopes that your Excellency will grant his request.[5]

As the council looked favorably upon Gauld's petition, a tract "on the Northeast side of the East Branch of Pearl River about seven leagues

---

4. C.O. 5/605: ff. 338–40. Robert R. Rea is indebted to Prof. Robin F. A. Fabel for bringing this entry to his attention.

5. The original petition in the Henry E. Huntington Library and Art Gallery, San Marino, California; microfilm in P. K. Yonge Library of Florida History, Gainesville.

The Surveyor at Home

from the Mouth" was surveyed November 7, 1776, by Elias Durnford and granted by Governor Chester on December 12.[6]

It is reasonable to think that George Gauld envisioned the eventual development of this property south of Picayune, Mississippi. His work on the coastal survey had been completed, for the most part, by 1776, and its interruption that year by the revolutionary disturbances to the north would have made him think seriously about the future. After a dozen years away from the British Isles, West Florida was his home, and there he might expect to live out his remaining years most comfortably. The Pearl River settlements were growing, and other prominent West Floridians were establishing plantations in the rich bottomlands. Perhaps here George Gauld intended to settle, establishing a new home in new surroundings and, in personal terms, under new circumstances.

Early in the year 1775, the bachelor surveyor took a wife. The exact date of his wedding cannot be established, but on January 31 the ubiquitous Philip Livingston, in his capacity as provincial secretary, issued a marriage license to George Gauld. It is to be hoped that the forty-three-year-old Gauld was not as derelict in his duties as young Livingston, who failed to record the receipt of the 6s. 5d. fee in his account book until July 15, when he cryptically explained that it had been "omitted being charged till now as I got the Bond signed and delivered the License at his new House."[7] Evidently, instead of issuing the license and receiving and recording the fee at his office, Livingston did Gauld the honor of personally delivering the document, and during the winter of 1774–75, in anticipation of his marriage, the surveyor had erected another dwelling on his town lot. There was ample room, and in all likelihood the bachelor's quarters, though adequate for his work and part-time residence, were unsuitable for the wife of a man of his standing. A more generous structure, with the roof extending over a modest porch or "piazza," as contemporaries termed it, would have been the style. Larger, not only because Gauld's household had increased, but also, perhaps, because the lady brought with her something more than mere personal belongings. It is not farfetched to think that the self-sufficient bachelor may have taken one of Pensacola's many widows as his wife. Unwed

---

6. C.O. 5/608: 228–29; Walter Lowrie, ed., *American State Papers* (Washington, D.C., 1834), 3:6, 19. John Payne also secured a 2,000-acre grant on the Pearl River at this time.

7. Provincial Secretary's Account Book "E," p. 53: West Florida Papers, Library of Congress.

single women were in scarce supply, widows plentiful, as the fevers and other illnesses of the subtropical zone swept through the town's population. Dr. Lorimer's best medicine was quinine ("Peruvian bark"), and he recognized that it was insufficient to save the lives of British soldiers and settlers who arrived from the colder climes just as the rigors of the summer were beginning. Quite possibly the lady was the widow of an army officer, a woman accustomed to ordering a household for a husband whose duty took him afar and for long periods. She would have needed those qualities, and Gauld would have appreciated them — but all is conjecture. Her name was Ann, and she was to survive George by many years, taking pride and finding some advantage in his cartographic achievements.

When he was at home in Pensacola, as he was three-fourths of the time after his marriage, Gauld would have welcomed and entertained a stream of visitors eager to have his professional advice and to exchange information of a geographic and scientific sort. Dr. John Lorimer and his wife were the Gaulds' most intimate friends.[8] Their common interests in natural science, in western lands, and in colonial government, and their many shared experiences, bound them together. Lieutenant Governor Elias Durnford and his lady belonged to a higher social category than the Gaulds, but Durnford's involvement in surveying and in the Iberville canal project would have drawn him to Gauld's side. Perhaps they recalled together that other army engineer, Lieutenant Philip Pittman, with whom Gauld had gone to Fort St. Marks in 1767. They might have read with interest the book that Pittman brought out in London in 1770, *The Present State of the European Settlements on the Mississippi*, and with knowledgeable eyes they would have studied the maps which it contained.[9] They knew, at least, that Pittman did his own surveying and owed no debt to others. That was not quite the case with a more recent visitor and cartographer.

The solitary wanderings of colonial geographers and naturalists made them dependent upon common friends and connections, as was

8. Gauld's intimacy with the Lorimers is specified by his will wherein both are set down to receive suits of mourning and rings. Gauld referred to Mrs. Lorimer's "kindness" and to a friendship with John that had subsisted since their early years when they were first acquainted — a phrase which might suggest that their familiarity antedated their coming to West Florida in 1764 and 1765. P.R.O. B 11/1091.

9. Robert R. Rea, Introduction to Pittman, *The Present State of the European Settlements on the Mississippi*, pp. xl–xlvi. There is also an edition by John Francis McDermott (Memphis State University Press, 1977).

true of George Gauld, Bernard Romans, and their mutual friend Dr. John Lorimer, all members of the American Philosophical Society. Romans was a Lowlander who was employed by the surveyor general of the Southern District, William Gerard De Brahm, as "Draughtsman, Mathematician, Navigator." He began working in the Florida peninsula in 1766, won appointment as deputy surveyor for Georgia, returned to Florida in 1769-70 as De Brahm's principal assistant, and was active on the west coast of the peninsula as far north as Tampa Bay. In 1770-71 he operated independently, exploring the Florida Gulf Coast north and west until, in August 1771, he reached Pensacola. He was there employed by Indian Superintendent John Stuart, who was currently concerned with drawing the southern Indian boundary line,[10] and won the favor of Governor Peter Chester for whom he did "some small affairs." To Chester, Romans appeared "to be an ingenious man & both a Naturalist & a Botanist." Thinking him "worthy of some Encouragement," Chester suggested to the home government that Romans be granted fifty or sixty pounds a year "in order to induce him to Continue in this Colony." Romans was consequently appointed provincial botanist in 1773, but he left West Florida in February of that year (evidently sailing from New Orleans) and made his way north to New York in search of a publisher for the material he had accumulated. By April 1775, Romans's *A Concise Natural History of East and West Florida*, with the accompanying map, "Part of the Province of East Florida," was published. At about the same time Romans turned rebel and joined the fighting in New England. In the course of 1779 or 1780 he was captured by the British and imprisoned until the end of the war. He died at sea in 1784, while returning to the United States.[11]

Although a brief nineteen months at most, Bernard Romans's West Florida sojourn enabled him to achieve cartographic fame, for his book was "the best written and most complete" published account of the Floridas[12] and his charts much the best in print—as he was quick to point out. Romans had little to say for Pensacola, a town of "about

---

10. Louis De Vorsey, Jr., *The Indian Boundary in the Southern Colonies, 1763-1775* (Chapel Hill, N.C., 1966), pp. 215-22; P. Lee Phillips, *Notes on the Life and Works of Bernard Romans*, ed. John D. Ware, facsimile ed. (Gainesville, 1975), pp. 29-30.

11. Ware's Introduction to Phillips, *Notes*, pp. xl-lxxxvi, 47; Rembert W. Patrick's Introduction to Bernard Romans, *A Concise Natural History of East and West Florida*, facsimile ed. (Gainesville, 1962), pp. xi-xxx.

12. Patrick, Introduction to Romans, p. xlvii.

an hundred and eighty houses . . . built in general in good taste, but of timber,"[13] but he enjoyed unusually good relations with a number of West Floridians. Among the subscribers to his book (which was dedicated to the royal agent John Ellis) were Governor Peter Chester and his secretary, Philip Livingston, General Frederick Haldimand and his adjutant, Captain Francis Hutcheson, Councilman Evan Jones, Attorney General Edmund Rush Wegg, and Dr. John Lorimer, sometime Speaker of the House.

There can be no doubt that the scientist-surgeon (who succeeded to Romans's position as provincial botanist in 1776) was delighted by the company of a visitor with Romans's professional interests and experience. George Gauld, who was at sea when Romans arrived and only returned after Romans had left for the interior, departed for Jamaica, late in 1771, and Lorimer found Romans to be a happy successor in Pensacola's small circle of intellectuals. The itinerant naturalist was no less interested in the information that Lorimer could provide, and that included the maps and charts which the naval surveyor had left behind. Lorimer made these available to Romans and won notice in the *Concise Natural History* as "my very worthy friend."[14] It was by no means a one-sided relationship, for Romans provided Lorimer with sketches and information regarding the upper reaches of the Chester River which the good doctor offered to Gauld in Jamaica.[15]

Romans was bitterly critical of several self-proclaimed authorities who had earlier treated the subject of Florida, and he rightly anticipated criticism of his own "directions to navigators," an eighty-nine-page Appendix in the *Concise Natural History*. He insisted that the directions were "entirely founded on my own experience" and declared that he was:

free of copying or compiling from prior works, except in that part lying west from the Mississippi, *which was taken from* French manuscript *draughts; and the shape of the coasts and bays eastward from* Pensacola,

13. Romans, *Concise Natural History*, p. 303.
14. Ibid., p. 334.
15. Lorimer to Gauld, August 13, 1772: American Philosophical Society Library. On the West Florida scientific community see Robert R. Rea, "The King's Agent for West Florida," *Alabama Review* 16 (1963): 141–53, and Robert R. Rea and Jack D. L. Holmes, "Dr. John Lorimer and the Natural Sciences in British West Florida," *Southern Humanities Review* 4 (1970): 363–72.

The Surveyor at Home

to Cape St. Blas, *which i have followed from draughts occasionally seen in the hands of Mr.* Stuart, *(the super-intendant) and my good friend Dr.* J. Lorimer, *which draughts i take to be the work of Mr.* Gauld, *a very able and accurate observer.*[16]

The precise extent of Romans's borrowing would be almost impossible to determine. His map actually covered both Floridas and parts of the Bahamas, Cuba, and Spanish Louisiana, ranging well beyond the scope of Gauld's surveys, but he did incorporate parts of Gauld's description of Espíritu Santo or Tampa Bay, based on the 1765 survey, in his Appendix.

Two manuscript maps attest further to Romans's borrowing. "A general map of West Florida," dated 1772 and prepared by Romans for Governor Chester, acknowledges that "the coast and shape of the bays in this map were taken from the accurate surveys of George Gauld."[17] The following year "A map of West Florida, part of Et: Florida. Georgia part of So: Carolina i[n]cluding . . . & Chactaw Chicasaw & Creek Nation to Augusitus & CharlesTown," compiled for John Stuart and thought to have been drawn by Joseph Purcell with notes by David Taitt, ascribed to George Gauld the "Sea Coast from the Mouths of the Mississippi to Appatacha [Apalachee]."[18] Romans was ready enough to grant Gauld the entrances to the Mississippi and the coast between Pensacola and Apalachee, but he insisted that "though i kept the shape, the distances are different from his; and are such as i have reason to believe nearer the mark," being based upon his own "long and tedious cruize . . . on this coast in 1771." Furthermore, Romans claimed, "As for the soundings, i never had his draughts long enough, to have an opportunity of perusing them, but they are such as i have obtained during the same tedious cruize."[19] Romans's disclaimer is not very persuasive. He could have had no doubt whatsoever that the charts he saw were Gauld's, and if he had time to take the "shape" of the coast from them, he certainly had time to copy Gauld's soundings. This is not to discredit Bernard

---

16. Romans, Appendix, pp. i–ii.

17. Phillips, *Notes*, pp. 47–48, 74–75.

18. Christian Brun, *Guide to the Manuscript Maps in the William L. Clements Library* (Ann Arbor, Mich., 1959), pp. 160–61; William P. Cumming, *The Southeast in Early Maps* (Princeton, 1958), pp. 50–51, 252. This was one of two maps Romans delivered to Gage; Phillips, *Notes*, pp. *lii, lx, xciv* n.78.

19. Romans, Appendix, p. ii.

G^{eo.} Gauld — Cartographer

Romans's undoubted contributions to cartography but to emphasize the fact that he was not a man to state the truth simply and plainly.

The accusation of plagiarism surfaced soon enough, and in exculpation Romans published a lengthy advertisement in the New York newspapers. It had been charged that his work was "a piracy obtained at Pensacola by the help of Dr. *John Lorimer.*" The truth was, Romans declared, he had completed two-thirds of his work "two years before I knew Pensacola, or had the honour of Dr. *Lorimer's* inestimable acquaintance." Far from pirating Gauld's charts with Lorimer's aid, at the doctor's request Romans had sent him "an extempore sketch of the coast of Florida" to be forwarded to Gauld in Jamaica and had recently dispatched a copy of his earlier work to Pensacola in order that the West Floridians might have full benefit of his cartography. Romans effusively described Lorimer and Gauld as connected in "the most strict, honourable and tender knot of friendship, perhaps existing on our globe, and whose acquaintance I esteem a valuable jewel." By repeated reference to Lorimer, and by implication, Romans sought to associate himself with that "tender knot," though he was less than generous toward Dr. Lorimer when he added that "neither of these two gentlemen pretend to be in the least acquainted with the science of Botany." His correspondence with Lorimer, published as evidence of their innocence of wrongdoing, indicates that the material Romans sent to Pensacola was "a rough draught of *East-Florida*" for Gauld's inspection. Lorimer, on his part, urged Romans to confine his publication to that area, leaving the western shores to Gauld; but the surgeon did propose to "ask Mr. *GAULD* what he can spare you for your better information relative to *West-Florida.*" Finally, Romans flatly stated that "the mouths of the Missisippi are exactly taken by Mr. *Gauld,* and inserted in my work by his permission, and I am now in expectation of that gentleman's assistance for the East part of West-Florida, which shall be duly acknowledged by me, and thus all stain or imputation of imposition I think is clearly removed."[20] Evidence of Gauld's acquiescence in Romans's borrowing is lacking, but subsequent publishers of Romans's work obviously felt that the naval surveyor's name lent credence to its accuracy. *The Complete Pilot for the Gulf Passage; or Directions for Sailing through the Gulf of Florida,* named also *New Bahama*

20. Phillips, *Notes,* pp. 27–30.

The Surveyor at Home

Channel and the Neighbouring Parts, first printed for Robert Sayer in London in 1789, added the names of both De Brahm and Gauld to the title page.

Whatever the reactions of John Lorimer and George Gauld to the Romans affair, they kept their opinions to themselves. Gauld was aware of the risks involved in the circulation of his manuscript charts and took what precautions he could. Haldimand, who sent one of Gauld's plans to General Gage in 1772, advised his superior that he had had to promise that no one would be allowed to copy it.[21] If John Lorimer's guard had been penetrated by the ebullience of Bernard Romans, George Gauld was not one to hold an excess of enthusiasm against his closest friend. As for Romans's betrayal of his loyalty to the crown, he was not the only erstwhile West Floridian to take that route.

Gauld's relations with a fourth mapmaker of the colonial frontier may have been more intimate in personal terms, certainly was in respect to their work, and offers a most striking example of the unacknowledged use of Gauld's materials. Thomas Hutchins was a military engineer employed by General Gage and twice a visitor in West Florida. In 1766, as an assistant to Captain Harry Gordon, Ensign Hutchins descended the Ohio River from Fort Pitt, stopped off at Fort Chartres where he encountered Philip Pittman, and proceeded to Pensacola from New Orleans. Gordon's expedition was intended merely as an inspection tour of the forts and posts in the newly acquired territories; he and Hutchins stayed in Pensacola only a few days, November 4-12, 1766 (at that time Gauld was in town, having been forced to curtail his activities by the great hurricane of October). In 1772, Hutchins returned to the colony and resided there until 1776. During this period he established an informal household in Pensacola and "entered into a minute examination" of the "coasts, harbours, lakes, and rivers" of the province. This, Hutchins boasted, "made me perfectly acquainted with their situation, bearings, soundings, and every particular requisite to be known by Navigators," for whose benefit he was subsequently "induced to make my observations public." In 1776, Thomas Hutchins returned to England. Three years later he was discovered to be in secret correspondence with Americans in Paris; he was arrested on charges of high treason but succeeded in fleeing the country and making his way to the United

21. Haldimand to Gage, May 14, 1772: Gage Papers.

States, where, with Benjamin Franklin's recommendation, he gained employment as geographer to the United States, a position he held from 1781 until his death in 1789. In 1784, Hutchins published *An Historical Narrative and Topographical Description of Louisiana, and West-Florida*, a work of some importance in Gulf Coast historiography.[22]

Inasmuch as both Gauld and Hutchins were resident in Pensacola at the same time and for a period of roughly two and a half years, some credence may be given to Hutchins's declaration that he "had the assistance of the remarks and surveys, so far as relates to the mouths of the Mississippi and the coast and soundings of West-Florida, of the late ingenious Mr. George Gauld, a Gentleman who was employed by the Lords of the British Admiralty for the express purpose of making an accurate chart of the abovementioned places." Hutchins noted that he "also had recourse, in describing some parts of the Mississippi, to the publication of Captain Pitman," whose work he had been shown in manuscript and which Harry Gordon judged to be "exacter than any thing we could do in tumbling down this rapid Torrent" of the Mississippi.[23]

Hutchins's debts were indeed great, far greater than he admitted. Comparison of Hutchins' *Historical Narrative* with Pittman's *Present State of the European Settlements* discloses that the former reprints, almost verbatim, lengthy passages from his predecessor's book.[24] Even more revealing is a comparison with Gauld's "General Description . . . of West Florida, 1769," which lay unpublished in the archives of the American Philosophical Society when Hutchins, as geographer of the United States, brought out his *Narrative*. Page after page of Gauld's work appears virtually unchanged in Hutchins's book. Modifications are of an editorial nature, or intended to hide the identity of the real author and the date at which he wrote, or consist of mere rearrangement of parts. The sum total of the unacknowledged material borrowed from Gauld alone fills more than a third of Hutchins's book. Only when he reproduced Gauld's "Remarks on the Tortugas" did Hutchins specify his debt to the naval surveyor, and then, it

---

22. Thomas Hutchins, *An Historical Narrative and Topographical Description of Louisiana, and West Florida*, Introduction by Joseph G. Tregle, Jr., 1784; Floridiana Facsimile & Reprint Series (Gainesville, 1968), Introduction, pp. *xxii–xxxv*, and preface, p. iii. The most extensive study of Hutchins is Anna M. Quattrocchi's "Thomas Hutchins, 1730–1789" (Ph.D. diss., University of Pittsburgh, 1944).

23. Ibid. See also Robert R. Rea, Introduction to Pittman, *The Present State*, p. xxxiii.

24. Tregle's Introduction to Hutchins, pp. xlii–xliii, states the case modestly.

might be assumed, because they had been published under Gauld's name prior to the appearance of Hutchins's book.

Hutchins's most recent annotator has suggested that an examination of the Hutchins manuscripts "gives a fairly clear but by no means complete view of the process by which the *Historical Narrative* was brought to publication." It appears that the thirty-three pages of Hutchins's book that are almost pure Gauld (though not so identified) were dispatched to Gage as a report entitled "General Description of the Sea Coasts, Harbours, Lakes, Rivers, etc. of the Province of West Florida" which Hutchins:

prepared during the early days of his assignment at Pensacola [i.e. ca. 1772–1773]. *Sometime during later years, interesting enough, he seems to have become confused as to the actual period of its first composition. The date 1769 may be noted affixed to the end of the title which heads it, a superscription obviously added subsequent to the writing of the body of the report.*

Two copies of this manuscript found in the Hutchins papers contain the editorial revisions made when Hutchins prepared the document for publication. "Any doubts raised by the peculiar '1769'," it has been suggested, may be eased by observing that the changes Hutchins inserted refer to his own activity in West Florida in 1772 and 1775.[25] Unhappily, quite a different conclusion is required by a comparison of Gauld's and Hutchins's works. Hutchins's unthinking reproduction of the date "1769," not to mention his exact use of Gauld's title for his report to Gage, proclaims his source. His careful removal of identifying phrases from verbatim passages proves his guilt. He copied Gauld's remarks on Espiritu Sancto, for example, up to a point; but he was forced to delete the damning words "which I survey'd in 1765"—a feat that only George Gauld could claim.[26] The geographer to the United States was a first-class plagiarist, but his victims were dead when he published, and Thomas Hutchins had emerged from the American Revolution on the winning side.

The victor customarily writes the history of human conflicts, but it is exceptionally painful when he does so with the pen of the van-

25. Ibid., pp. xl–xli.
26. Comparison of the Hutchins Papers at the Pennsylvania Historical Society and the Gauld manuscript at the American Philosophical Society Library, Philadelphia, does not bear out Tregle's contention but proves the charge of plagiarism. For the example cited, see Hutchins, p. 89; Gauld, "A General Description of West Florida, 1769," p. 29.

quished. Justice requires that a considerable measure of long overdue credit now be given to George Gauld. When and how his "General Description" passed into Hutchins's hands may not be known, but it appears most likely that it had done so before Gauld returned to West Florida in the fall of 1773. Again, one is forced back to the understandable desire of Dr. John Lorimer to give the widest circulation to whatever scientific and geographic knowledge of West Florida could be found, a proper scientific attitude devoid of calculation and innocent of the devious uses to which other men might put the hard-won achievements of his friend George Gauld. Happily, the naval surveyor did not live to see Hutchins's flaunting of the prize, though he probably learned of his treason and desertion when he returned to London at the end of the American Revolution.

CHAPTER XII

The Impact of the American Revolution

The infant colony of West Florida, lying on the frontier of British North America, scarcely felt the tremors that heralded the American Revolution. Its sparsely populated lands and struggling towns bred neither minutemen nor mobs; its legislature might oppose but could not control a royal administration that was financially independent of colonial taxation; the visible presence of potentially dangerous Indian tribes justified the maintenance of military forces sufficient to discourage violent protest had West Floridians not recognized in the redcoats an economic boon to their well-being as well as a guarantee of their security; and in contrast to any other British colony in North America, West Florida faced an armed and hostile neighbor in Spanish Louisiana. The preconditions of revolution did not exist in West Florida—not that the ground was infertile or the seed dormant; but the tree of liberty had yet to mature on the Gulf Coast.

West Florida's weakness guaranteed her loyalty; her isolation from the disturbances to the north made her appear an attractive haven for loyalists driven by either their fears or very real persecution from nearly every colony in which rebellion raised its head. Loyalist migration swelled the population of West Florida to perhaps twice its prerevolutionary size,[1] but such growth was not an unmixed blessing. Some of the immigrants were men of property, able to purchase new land and contribute to the colonial economy. Others fled the Revolution with little save their loyalty to king and country to support them in an unfamiliar environment. They crowded into Pensacola where they found little shelter and less opportunity to make new lives for themselves. Their needs contributed to the inflation of prices and a persistent shortage of food and goods. Nor were the newly arrived loyalists the strongest of reinforcements. Some left their homes because they would not fight; others fled in defeat; all looked to their king to protect them rather than to their own resources.

1. J. Barton Starr, *Tories, Dons, and Rebels: The American Revolution in British West Florida* (Gainesville, 1976), pp. 230–31. See also J. Leitch Wright, *Florida in the American Revolution* (Gainesville, 1975).

G^{eo.} Gauld — Cartographer

The colony's defenses on land and sea were never impressive. The 16th Regiment, commanded by Lieutenant Colonel Alexander Dickson, was far below strength, and its veterans were debilitated by long service (since 1770) in an exhausting climate. The men of the 60th Regiment, sent over from Jamaica in 1776 and 1778, were an unimpressive collection of raw recruits scarcely trained to handle their muskets. At the beginning of 1779, the 3rd Waldeck Regiment arrived, German mercenaries who lacked any interest in the cause for which they had been sold and sent to fight. These, with a few loyalist troops from Pennsylvania and Maryland, made up the military force at the disposal of Brigadier General John Campbell, to whom the defense of West Florida was entrusted in 1778.

The strong points of the colony were Manchac and Natchez on the Mississippi, Mobile, and Pensacola. They were far from adequate for any serious military purpose. Fort Bute, abandoned in 1768, was a ruin. Natchez's Fort Panmure was impressively sited but located at an indefensible extremity of the province. The finest fortification on the Gulf Coast was Fort Charlotte at Mobile. Constructed of brick by French engineers sixty years before the American Revolution, it was a modest gem of European military architecture — on paper. In fact, it was tumbling down from age and disrepair; but for the cost, it would have been dismantled and the brick shipped to Pensacola. The capital of West Florida was defended by the old stockade fort in the center of town that had never been more than a refuge in case of an Indian raid. To scatter the few available troops among these isolated posts would be to offer hostages to fortune and weaken every place in the process. The provincial government, headed by Peter Chester, was determined to preserve its little strength for the defense of Pensacola, where its members made their homes and conducted their public and private business. Their decision was based on selfish considerations, but it was reasonable from a military point of view. Pensacola would be secure, and if war came, Great Britain would doubtless send an army to seize New Orleans, as had been contemplated as recently as 1771. There were American rebels on the Mississippi River from time to time, but they were far beyond Governor Chester's reach and enjoyed the benevolent protection of the Spanish authorities in any case. Wild rumors of the imminent descent of an American army from Fort Pitt might cause a shudder of fear to run through the colony, but none came, and loyal settlers and Indian allies could not maintain a constant vigil on the long riparian frontier.

The Impact of the American Revolution

The proper defense of the colony lay at sea. From its inception until history disclosed its fate, West Florida's first and best protection was the navy; on that, every military authority was and had always been agreed. Only European military forces could reduce Mobile and Pensacola, and they must pass the guard of a fleet which, in 1763, truly ruled the waves. The ships — all too often the same ships — were there in 1776, and so too were some of the men; but few outside the navy thought what toll the years had taken or that bright young lieutenants — George Johnstone, for instance — might not become great admirals. Worse, no one responsible for Britain's war effort in 1776 cared to contemplate what the suppression of the American revolt would require when the rebels were openly abetted by both France and Spain. Consequently, there were never enough ships, enough men, enough supplies to maintain the necessary naval strength in any theater of war at all times, nor, on such a distant post as the Jamaica Station, at any time.

Armed rebellion began at Lexington and Concord in 1775; Gage held Boston with Haldimand at his side, until that bloody day on Bunker Hill brought his retirement, and Howe was forced to withdraw in 1776. American independence had hardly been proclaimed when Howe returned with the greatest invasion army Britain had ever raised and seized New York. Philadelphia fell to him in 1777, although Burgoyne's luck ran out at Saratoga that same year.

Far to the south, at Pensacola, life went on as usual. George Gauld's surveying was uninterrupted by the preliminaries, but in 1776 the increased pace of action kept him close to home. American privateers were not to be trifled with in a surveying sloop, "armed" though she might be, and Gauld became exceedingly cautious about sending charts to Jamaica by any save naval vessels in succeeding years. The hiatus of 1776 could not have been entirely unwelcome to the newly married surveyor, but in 1777 he was eager to renew his work, and as if anticipating the future he extended his reach to the most distant Spanish shore that time allowed.

George Gauld's last voyage as a surveyor coincided with the elevation of Bernardo de Gálvez to the governorship of Spanish Louisiana. In 1777 Gálvez demonstrated that he possessed not only the martial prowess expected of a young hidalgo but equal daring in diplomacy. His predecessor had implemented the Spanish government's policy of secretly aiding the American cause by supporting the rebel agent Oliver Pollock (once a West Floridian), but in April, Gál-

G^{eo.} Gauld — Cartographer

vez directly attacked British mercantile interests by seizing eleven vessels at New Orleans (including that belonging to one William Pickles) and banishing British merchants from Spanish territory. Gauld saw something of this affair when *Florida* was anchored at the mouth of the Mississippi. British reaction was prompt, but British protests were turned aside by Latin hyperbole. Neither side wanted war; neither was prepared for hostilities. Gálvez continued to extend every possible assistance to the rebel cause, and little British cruisers, such as *West Florida*, did what they could to intercept Spanish and American smugglers in the lakes and coastal waters.

While Gauld was perfecting his chart of the Gulf Coast west of the Mississippi, former West Floridian James Willing was preparing to attack the colony. Floating down the Ohio and Mississippi rivers in January 1778, Willing and some hundred ruffians appeared at Walnut Hills, Natchez, and Manchac between February 18 and 23. Enjoying complete surprise and overwhelming force, they soon turned from patriotic conquest to profitable piracy, and in New Orleans they found a market for their loot and a sanctuary from British reprisals.

News of the Willing raid reached Pensacola March 4, and the navy went into action. H. M. S. *Sylph*, under Captain John Fergusson, raced to New Orleans and was soon followed by Captain Joseph Nunn in *Hound*. With rebel forces hospitably ensconced in New Orleans, and British guns trained on the town, Bernardo de Gálvez had reason to display his best diplomatic talent. He had little doubt of what would happen if *Hound* and *Sylph* loosed their broadsides on New Orleans, for the city's defenses looked away from the river, not toward it. Happily for Gálvez, Fergusson and Nunn understood that their mission was to recover as much British property as possible, not to start a war. With that in mind, they restrained their tempers and their gunners. Gálvez made timely restitution of sufficient stolen property to persuade the British refugees that more might follow, and with the blessing of the deluded merchants and planters of Manchac and Natchez the British naval force withdrew and returned to Pensacola.[2]

All thoughts of coastal surveying disappeared for George Gauld

---

2. Ibid., pp. 78–105; Elizabeth Jones Conover, "British West Florida's Mississippi Frontier during the American Revolution" (master's thesis, Auburn University, 1972), pp. 58–105, summarized in her "British West Florida's Mississippi Frontier Posts, 1763–1769," *Alabama Review* 29 (1976): 177–207. Capt. Joseph Nunn to Parker, May 9, 1778: ADM 1/241.

The Impact of the American Revolution

with the news of Willing's raid. Local waters were no longer safe, for rumor quickly transported Willing's men from Manchac to Mobile and provided them with ships and guns. At the moment, Gauld was awaiting the arrival of his new surveying schooner, but no sooner did she enter Pensacola harbor than she was ordered westward in search of rebel privateers. The surveyor remained behind and left such pursuits to fighting men. Two other West Floridians, Adam Chrystie and Anthony Hutchins (the mapmaker's brother), took matters in their own hands and enjoyed a bloody revenge upon those of Willing's cohorts who remained at Manchac and Natchez. Their daring and success made them heroes among their fellow colonists, and Gauld would see more of them.[3]

Willing's raid stirred West Florida to the core. Frantic appeals from Peter Chester to Sir Peter Parker drew what support the harassed and shorthanded Jamaica Station could provide. Merchants on the Mississippi demanded protection, and in April the restoration of a military post at Manchac was begun. In January 1779, Brigadier Campbell and the Waldeck Regiment arrived. The British government and army had provided all the manpower it could afford. The rest was up to West Florida whose population, swollen by refugees, might produce a considerable militia if a General Assembly would create one, and its citizens, highly critical of their governor's feeble gestures, demanded that the necessary steps be taken.

Peter Chester was no lover of assemblies. In 1772 he had refused to allow an assembly to convene because Mobile's voters insisted upon limiting the terms of their representatives, and Chester insisted that the determination of the life of an assembly was a prerogative of the governor. In 1777 he ignored instructions from Whitehall to convene an assembly in order to regulate the Indian trade, but obvious danger and widespread condemnation of his failure to preserve the security of the western parts of the colony forced him to acquiesce in 1778. On April 27, Chester ordered the issuance of election writs, but because of the disturbed state of the Mississippi settlements, to whom he gave representation in the Commons, the assembly did not gather until October 1.

The General Assembly of 1778 contained the largest number of representatives ever summoned to Pensacola, but two-thirds of its members were novices; only two had more experience than the se-

---

3. Starr, *Tories, Dons, and Rebels*, pp. 103–5, 108–12.

nior member for the district of Pensacola, George Gauld. The Commons House differed further from its predecessors, for in an effort to dominate Mobile, Chester had thoroughly revised the representative districts, a gerrymander that appeared to double the number of members from Pensacola (or halve those from Mobile) and added eight new seats for Manchac and Natchez. Chester's argument that the west deserved a legislative voice was sound, but the reason for his benevolence was all too evident. Had he met the assembly in April, he might have fared better, but by October his critics had had time to organize; the Commons came to Pensacola prepared to seize the initiative. Among them sat the current heroes of the Willing affair, Adam Chrystie and Anthony Hutchins; but the men of action would take opposite sides in the ensuing disputes, thereby dividing the house and bringing to prominence a moderate middle party led by George Gauld.

The Commons gathered on October 1 and immediately singled out Gauld, Chrystie, and Hutchins to notify the governor of their presence. In his speech to the two houses, Chester insisted that during the Willing episode he and his administration had taken "every measure" that had been thought "conducive to His Majesty's service." He warned of the probability of war with France (a fact six months before he spoke) and recommended the passage of "an Act for the establishing and proper regulating of a militia which, as the Province increases in numbers, must add greatly to our safety and defense." In conclusion he flattered the Commons on their "real wishes to promote the public welfare" and urged them "to cultivate union and harmony between the different branches of the Legislature."

As the Commons settled into the routine of business, George Gauld was assigned to the committee to draft a reply to the governor's speech, to the Rules Committee, and to that on Privileges and Elections. He supported the votes of thanks extended to Captains Nunn and Fergusson, Adam Chrystie, and Anthony Hutchins for their gallant efforts during the recent troubles on the Mississippi and participated in the deliberations that produced a reply to Governor Chester's opening speech. That message politely but firmly chastized the governor for being dilatory in sending assistance to Manchac and Natchez but promised cooperation in all matters to which he had directed their attention. With the formalities completed, John Miller moved that the Committee on Privileges and Elections consider the apparent omission of representation from the town of Mobile,

The Impact of the American Revolution

lumped in Chester's election writ with the surrounding district of Mobile, whereas Pensacola enjoyed both town and district representation. The Elections Committee was not prepared to report, as directed, on October 6, but the house did extend its thanks to "Mr. John Payne of His Majesty's Navy for the services he has rendered to this province on various occasions for several years past." It may be assumed that George Gauld took the lead in honoring his old comrade and assistant.

The Privileges and Elections report was made on Wednesday, October 7, and delivered to the house by John Miller. It noted the relative reduction of representation from Mobile, but Miller's attempt to persuade the house to demand an immediate explanation from the governor was defeated. The house appointed a committee to draft a militia bill, and Gauld was put on another committee to prepare a bill for attaching the property of absent debtors.

The house did no business on the ninth of October because of the devastating hurricane that battered Pensacola that night. George Gauld had experienced Gulf Coast storms before, but no one had seen "such irresistible fury and violence" as this. The Pensacola waterfront was swept away: wharves, stores, houses, and batteries. Every ship in the bay except H.M.S. *Sylph* and *Stork* was sunk, driven ashore, or, like *Hound*, reduced to a mass of tangled masts, spars, and cordage. Property damage was tremendous, though, amazingly, no more than seven lives were reported lost. For the next week the whole community was fully occupied in cleaning up and assessing the destruction. Months later, Pensacola would still be a scene of "desolation and ruin."[4]

On October 15, the Commons resumed their session. George Gauld rose to move for a committee to draw up a bill "for Ascertaining the Number of Representatives, regulating Elections, etc.," and a committee was duly established. Gauld's motion demonstrated his sympathy with the dissidents in the house, but in contrast to John Miller (who was also a member of the drafting committee), Gauld chose to proceed by legislative means and thereby avoid a confronta-

---

4. Ibid., p. 133; Lieut. Col. William Stiell to Germain, October 15, 1778: C.O. 5/595; Nunn to Parker, December 4, 1778: ADM 1/241.

Gauld had completed the formal draft of his 1777 survey and consigned it to Nunn's care when the hurricane struck. As *Hound* was unable to leave Pensacola for several months, Gauld revised and improved it before sending it on to Parker. Gauld to Parker, January 14, 1779: ADM 1/241.

tion with Governor Chester. That same day, Hutchins's militia bill was reported back and received its first reading. The critical issues of this General Assembly had begun to merge; on the nineteenth the house postponed the progress of the militia bill, and the next day it agreed to ask Chester for an explanation of the Mobile gerrymander. Gauld was given that task, along with Miller, and further demonstrated his concern by moving the house to inquire into other aspects of Chester's reapportionment of the legislature.

Three days later, Gauld reported the elections bill to the house, and the Bill for Establishing the Number of Representatives received its first reading. A motion for an immediate second reading was defeated, and the Commons received a message from the governor warning them against interfering with his prerogative. The next day the Commons took the governor's message under consideration and, with Gauld's support, assigned it to a committee of which the surveyor was a member. Its own elections bill was passed on October 23, and on Saturday, the twenty-fourth, when the engrossed bill was carried to the Upper House, the Commons proceeded with their consideration of the militia bill.

George Gauld had taken a prominent role in the passage of the elections bill, thereby placing himself firmly among the opponents of Governor Chester. He now swung his influence and vote to the side of those pressing for the completion of the militia bill. He clearly felt that having expressed their views on the popular side of the constitutional issue, the house should demonstrate good faith and responsibility by bringing a militia into being. Debate was interrupted by a request from the Upper House that a meeting might be held on the elections bill, and Gauld was routinely appointed to the conference committee which met on Wednesday, October 28.

The meeting of Commons and councilors at five o'clock Wednesday evening was the first of a series of efforts by moderate leaders in both houses to find grounds on which the interests of all parties might be satisfied. Gauld reported to the house on October 29 that the members of the council sitting with their committee had agreed that they should jointly address the governor requesting that Mobile might have its accustomed representative parity. This tactic was approved by the house, and at nine o'clock on Saturday, Gauld led a delegation into further conference with the committee of the Upper House. At that meeting the moderates in the Commons promised that if Chester would satisfy their demands regarding Mobile, they

209

would guarantee the passage of the militia measure, which was ready for its final reading. To their surprise and regret, the councilors produced a written response declaring that they were informed "from certain authority," obviously the governor, that the proposed course of action would not gain Chester's approval. As a sop, the councilors suggested that if the Commons memorialized the governor in suitably humble terms, he would forward their message to England. The Upper House tried to explain their refusal to undertake joint action by suggesting that it was improper for them to act in anything "respecting the privileges of the people." Gauld and the Commons committee replied — with some warmth, no doubt — that they were "well aware of the impropriety of it . . . yet we thought ourselves justifiable on the present occasion to try every possible means to get over the embarrassment we are involved in, that we might proceed to business with that unanimity which is necessary at all times, but more especially the present when perhaps the very existence of the Province is at stake."

Gauld and his colleagues had done everything possible to secure the electoral rights of West Florida's second city and the representative interests of the people at large. They had proceeded with the greatest decorum and propriety and had demonstrated their readiness to provide the colonial militia so desperately needed for the defense of the colony — if Governor Chester would withdraw from his spiteful attitude toward Mobile. Their hopes struck on the rock of gubernatorial pride and prerogative; on that all too visible hazard the ship of state soon foundered. On November 5, both houses were summoned to attend the governor in the Council Chamber, and there Peter Chester put an end to the General Assembly, accusing the Commons of having "matters of privileges more in view than His Majesty's interests and the . . . defense of the Colony." He adjourned the session until the following September, but he would never again allow a voice to the people's representatives in West Florida.[5]

George Gauld, a man without great political ambition but possessed of a strong sense of duty, was an experienced politician and the friend of others. No partisan, he was entrusted by his fellows with parliamentary functions that reflected the respect of Commons, council, and governor because he enjoyed the public virtues of honor, integrity, and fairness. From the first, Gauld placed himself

5. Robert R. Rea and Milo B. Howard, *The Minutes, Journals, and Acts of the General Assembly of British West Florida*, pp. 274–313.

on the side of popular representative government and opposed Peter Chester's petty chicanery with regard to the distribution of seats in the Commons. One may sense in some members a certain vindictiveness toward the governor; George Gauld appears to have acted on principle. With his friend and attorney, John Mitchell, he supported the militia bill, whereas the rest of the popular opposition party stubbornly refused to show the least conciliatory attitude toward the governor. Through the bare words of the last conference committee report, which he delivered, one can sense Gauld's disappointment, frustration, even a feeling of betrayal that all should fail because of one man's folly. Gauld occupied the unenviable ground of a moderate, and no doubt he suffered the jibes of the irreconcilables.

Governor Peter Chester was quick to single out his enemies, but he did not indict George Gauld among the parties responsible for the assembly's failure. Chester blamed Lieutenant Governor Elias Durnford, Speaker Adam Chrystie, and the Indian traders of Mobile. That so prominent a figure among the opposition as George Gauld should escape his vituperative pen is surprising until the naval surveyor's role in the dispute is properly recognized; then the governor's silence becomes an unspoken tribute to a truly distinguished West Floridian. George Gauld was one of the few opposition members of the assembly, one of the few prominent Pensacolans, who did not lend his name to the equally vicious attacks against Chester that were dispatched to London following the prorogation of the assembly.[6]

If the refusal of the General Assembly to provide a militia for colonial defense caused concern in Pensacola, it was soon eased by the arrival of the Waldeck Regiment and Brigadier General John Campbell. Neither the stolid Germans in their fancy uniforms nor the fussy British general took much comfort from their assignment. At the beginning of 1779, Pensacola was still recovering from the battering of the October hurricane; housing was inadequate, supplies low, and prices high. Between lengthy epistles to Whitehall explaining his problems and his fears, Campbell began to scatter his troops about the province and to construct costly fortifications at Pensacola. Naval vessels contributed their share of manpower and expertise for the battery at the Red Cliffs that was intended to seal Pensacola Bay against an invader, and George Gauld may have participated in such engineering activities. Admiral Parker's good intentions notwith-

6. See Lucille Griffith, "Peter Chester and the End of the British Empire in West Florida," *Alabama Review* 30 (1977).

standing, he was without a surveying vessel, and his friend John Payne was now captain of *West Florida*.

Months of consequential inactivity passed. Summer settled heavily upon Pensacola, and fevers and dysentery began to take their toll of Campbell's troops. Unknown to West Florida, Spain declared war upon Great Britain and sped word to New Orleans, where Bernardo de Gálvez waited expectantly. Forewarned and forearmed, on August 27 the brilliant young Spaniard launched the first blow as war came to West Florida.

Britain's first line of defense was at sea — the little sloop *West Florida*, commanded by Lieutenant John Payne, patrolling the waters of Lake Pontchartrain. On August 29, Payne sent his yawl with an officer and seven men toward Manchac to make contact with Lieutenant Colonel Dickson. The boat party was captured by Spaniards that same day, and Payne cruised the lake, awaiting their return, until September 10, quite unaware of the state of war. At two o'clock that afternoon he sighted a strange sail and gave chase. Coming alongside and hailing the stranger, Payne was informed that she was a merchant vessel bound from Pensacola with supplies for Manchac; but even as her captain spoke, grappling lines were thrown into the starboard quarter of *West Florida* and she began to receive fire. As luck would have it, the previous day Payne had removed the barricade on that side of the ship, and the sloop's deck lay open to the enemy's small arms. Payne had only fifteen men, and some of them were sickly; but he made a fight of it at odds of better than four to one against him. Twice the attackers tried to board *West Florida* and were thrust back. The third time they came in overwhelming numbers. John Payne received a mortal wound; his master's mate, Gerrald Savage, was also wounded and one man killed before the pitiful remainder threw down their weapons. The victor proved to be the American privateer *Morris* commanded by William Pickles, once a British merchant skipper, now an American naval officer serving with the Spaniards. *Morris* mounted eight guns and carried sixty-five men; her casualties — eight killed and a number wounded — are proof enough that Payne, Savage, and their baker's dozen fought gallantly.[7] Ironically, British naval orders to commence hostilities had reached Pensacola on September 8, and an eight-man boat had been dis-

---

7. Gerrald Savage to Captain Francis LeMontais, New Orleans, October 24, 1779: ADM 1/242, provides the only British eyewitness account. *An Account of The Surveys of Florida*, p. 26, presents the story as Gauld received it.

patched to the west to warn Payne, but lost time could not be recovered. Like *West Florida*, this messenger also fell into Spanish hands.[8]

Gálvez's success was complete. Manchac was overrun; Dickson and 350 troops capitulated at Baton Rouge on September 21, and Natchez was surrendered at the same time.[9] At a stroke, British West Florida was reduced to the Gulf ports of Mobile and Pensacola.

All but the last battle had been fought when Peter Chester formally announced on September 10 that a state of war existed, but the news had already swept through Pensacola. A delegation, headed by George Gauld and including John Mitchell, came before the council that very day with a petition from forty-eight citizens praying that they might be embodied into a temporary militia and serve under their own elected officers. The council thanked them for their "Loyalty and Zeal" and granted their petition. The home defense force which Gauld helped to organize comprised two volunteer companies by March 1780. In addition, the citizenry offered bounties to any who would enroll in the regular services, offered to charter civilian vessels for military use, and constructed a redoubt at the east end of Pensacola. When recounting their contribution to the defense of Pensacola they noted that the place could produce no more than 107 men capable of bearing arms (excluding government officials); forty of these were employed in the king's works — the fortifications on Gage Hill and at the Red Cliffs — and the rest were enrolled in the volunteer companies.[10] These were not just summer soldiers, nor were they the riffraff of the town; they were solid citizens, property owners, and businessmen whose stake in the future of West Florida was great indeed.

For all, the approach of war brought new demands upon their competency. Gauld's civilian occupation naturally led to his drafting of military plans and maps, at least one being sent to headquarters at New York.[11] The naval surveyor was also consulted regarding communication with Manchac, and in spite of the persistence of wild schemes to canalize the upper reaches of the Iberville, Gauld finally

8. LeMontais to Parker, December 20, 1779: ADM 1/242.
9. Jack D.L. Holmes, *The 1779 "Marcha de Gálvez"* (Baton Rouge, [1974]), and *Honor and Fidelity* (Birmingham, 1965); Robert V. Haynes, *The Natchez District and the American Revolution* (Jackson, 1976).
10. Council minutes, September 10, 1779, March 3, 1780; C.O. 5/635.
11. "Plan of Manchac, 1774," drafted for Sir Henry Clinton at Gen. Campbell's request, 1779. Clinton Papers, William L. Clements Library, Ann Arbor. Brun, p. 165.

persuaded General Campbell of the futility of the project. In fact, the surveyor rightly judged that the land on which Fort Bute once stood, on which the Manchac post was located in 1779, would eventually be washed away by the Mississippi River.[12] Gauld's knowledge of West Florida waters and his naval associations particularly recommended him for other special services.

In January 1777, the justices of the peace at Pensacola arrested John Webley, master of the schooner *Sally,* and his crew on suspicion of their having plundered the brigantine *William and Elizabeth,* which had been stranded on the Chandeleurs. Examination of Webley disclosed that several other merchant captains had also looted the stricken ship and were selling the stolen cargo in Pensacola. The justices promptly ordered wholesale arrests of the accused and committed them to trial or bound them over to give evidence. Such vigorous law enforcement so reduced the number of available seamen at Pensacola that hardly a ship was able to leave harbor. As the regular session of the Court of Oyer and Terminer was not scheduled until the second Tuesday in April, the justices asked the council to appoint a special court to try the accused so that witnesses and innocent parties might be released. Seeking to establish the proper judicial authority in the case, the council sent for George Gauld and asked him to determine whether or not the wreck of *William and Elizabeth* lay within the jurisdiction of the colony and His Majesty's dominions. The naval surveyor consulted his charts and advised the council that as the derelict lay seventeen leagues off the mainland coast it was not within the provincial boundaries, but it was clearly within the limits of His Majesty's dominions. Attorney General Edmund Rush Wegg therefore held that the case should be heard in the West Florida Court of Oyer and Terminer, and a special commission was issued for the trial of Webley and his fellow looters.[13]

In the winter of 1780-81, Gauld was involved in a far more intriguing legal squabble. Navy vessels based at Pensacola were as busy intercepting Spanish coastal shipping as winds, weather, and the demands of Governor Chester and General Campbell would permit. Prizes were frequently taken, and to the officers of the Royal Navy they represented the tangible rewards of service. Captain Robert Deans of *Mentor* and Lieutenants James McNemara of *Hound* and Timothy Kelly of *Port Royal* depended upon agents to manage the

12. Campbell to Germain, February 10, 1779: C.O. 5/597.
13. Council minutes, January 28, 1777: C.O. 5/634.

legal aspects of securing their share of prize money when they brought Spanish ships into Pensacola, and George Gauld and his friend John Mitchell acted in that capacity. It was the agents' duty to present to the local vice-admiralty court, in which Elihu Hall Bay sat as judge, the libel or right of the naval officers to each prize. When awarded by the court, the prize and its cargo was appraised and put up for sale at public auction, the officers and crew of the captor receiving a portion after posting surety that the sale was legal. Appearance before the bench was restricted to certain lawyers, however, and at Pensacola, Attorney General Edmund Rush Wegg enjoyed a virtual monopoly of admiralty court business. Wegg was perhaps no worse than most prize court lawyers, but he had a reputation for demanding exorbitant fees, and with Judge Bay's connivance he could force the naval officers to employ his services and pay his fees.[14] Although a common situation, it was one that did not give much satisfaction to harried naval commanders who knew that a major Spanish assault on Pensacola was imminent.

In fact, the case of *El Grand Poder de Dios* arose as a result of the Spanish invasion attempt of October 1780, which was disrupted by a hurricane. The ill-fated Spanish supply vessel, unarmed, dismasted, and laden with powder and artillery supplies, was taken on November 15 by the merchant sloop *Nelly*, John Buttermire master, while drifting helplessly off Mobile bar. As Buttermire held no letter of marque he could make no legal claim to such a capture, and *El Grand Poder* would automatically pass into the custody of the navy if he towed her into Pensacola harbor. Buttermire proposed to his mate, William Gosling, that they simply plunder the Spaniard, transfer the cargo to *Nelly*, dispose of the crew, and set her adrift — enjoying their profits in silence. Gosling refused to participate in such a conspiracy, and on November 17, four leagues off the Red Cliffs, the Spanish ship was boarded by boat parties from H. M. S. *Port Royal* and *Hound* and brought in as a navy prize.

Lieutenant McNemara and his senior, Captain Deans, directed their agents, Gauld and Mitchell, to get *El Grand Poder* libeled in the admiralty court on November 19, and the agents applied to Edmund Rush Wegg to act as their proctor, to which he agreed. John Buttermire was understandably disgruntled at being foiled in his piratical intentions and seems to have found a sympathizer in the deputy re-

14. Gauld and Mitchell to Deans, September 18, 1780: ADM 1/1709; see Carl Ubbelhode, *The Vice-Admiralty Courts and the American Revolution* (Chapel Hill, N.C., 1960), pp. 5–14.

ceiver general of the droits of the Admiralty, who entered a counterclaim to *El Grand Poder* and persuaded Wegg to represent him in the case without informing Gauld and Mitchell that he had switched clients. Rumor soon unveiled Wegg's treachery, however, and at ten o'clock Monday morning, November 27, George Gauld and John Mitchell confronted the attorney general in his office and demanded an explanation. Wegg admitted that on the preceding Friday he had agreed to represent the deputy receiver general, and he grudgingly conceded that under the circumstances Gauld and Mitchell ought to be permitted to employ another lawyer (a dispensation, Gauld observed, that Wegg had refused in a previous case). At that point James McNemara was convinced that the profits of the prize would all be pocketed by the lawyers and the court. The agents proceeded as best they could, engaging the lawyers Morrison and John Bay (probably the son of the admiralty court judge) to represent them, but because of the deputy receiver general's intervention, the naval captains now stood in the position of claimants or defendants, rather than libelants, and a new set of documents had to be drawn up before their case could be presented.

Trial was set for December 8, but Wegg was not ready to offer his arguments that day, so the hearing was put off until the eleventh, when the wheels of justice began to creak. There being no question in law that Buttermire's lack of letters of marque gave the naval officers a clear claim to *El Grand Poder,* the facts were established by the fifteenth, and the agents were then prepared to complete the transaction, having a civilian appraiser and the required surety in court. Wegg, however, proposed a different appraiser (an official of the Army Ordnance Department, thereby introducing a nice touch of service rivalry) and succeeded in delaying further action until December 21. Gauld and Mitchell secured their own army appraiser to evaluate the prize, but the final decision was presented to the court by the provost marshal (yet another provincial bureaucrat to be paid) who set *El Grand Poder*'s worth at £7,844. This sum was considerably less than Gauld and Mitchell expected. Artillery Captain Johnston had valued the cargo of powder alone at £3,000–4,000 but the marshal returned it at a mere £1,869.13.4, and from that, £335.8.5 in court costs were yet to be deducted.

Final settlement was now delayed by the Christmas holiday. On December 30, concluding arguments were heard. To the navy's surprise, Councilor Wegg quite ignored the interests of his client, the deputy receiver general, and pleaded those of Captain Buttermire and "his gallant crew." Captain Robert Deans was absolutely infuriated. In his opinion, Buttermire deserved to be arrested and charged with robbery, for the Spanish captives had complained of receiving harsh treatment at Buttermire's hands, but Attorney General Edmund Rush Wegg had refused to pursue the matter when Deans brought it before him.

Admiralty Judge Elihu Hall Bay convened his court on January 2, 1781, and rendered what Gauld and Mitchell bitterly characterized as an "indecisive decision." Bay declared that Buttermire deserved the prize but might not have it because he had never claimed it. Neither had the naval officers established a right to the prize, for they had failed to produce in court the Act of Parliament establishing their claim in such a case — and the Vice-Admiralty judge refused to recognize naval custom as legally compelling. *El Grand Poder* was, therefore, ordered to be sold, the proceeds to be deposited with the Registry of the Court until a final determination should be rendered by the crown. As Gauld and Mitchell remarked, the navy officers had been cheated of their just deserts by their own Admiralty court.[15]

The loss of the Mississippi in 1779 threw British Pensacola upon the defensive, a familiar attitude for both Major General John Campbell and Governor Peter Chester; they did little but await the worst, and neither had many doubts that it would come. Unsure where Gálvez would strike next, they entrusted the defense of Mobile to Lieutenant Governor Elias Durnford and gave him a token force with which to hold a crumbling fortification. In January 1780, Gálvez embarked an expeditionary force at New Orleans, and by the first of March he had landed and was ready to attack Fort Charlotte. Two weeks later Durnford surrendered while a British relief column splashed about in the swamps at the head of the bay. Sudden Spanish success was followed by months of inactivity, however, for Pensacola could only be approached by sea, and Spanish commanders at Havana, fearful of British naval prowess, would not press an attack, imagining lightly armed merchant vessels in the harbor to be ships of the line.

15. Deans to Chester, November 30, 1780; Deans to Parker, February 15, 1781; Gauld and Mitchell, "Journal of the Proceedings": ADM 1/1709. Bay to Chester, February 21, 1781; Chester to Germain, February 22, 1781; Vice-Admiralty Court libel, January 2, 1781: C.O. 5/596. A somewhat different account of the case appears in Johnson, *British West Florida*, p. 231.

In fact, when *Hound* (20) and *Port Royal* (18) arrived in April 1780, Lieutenant McNemara found "no naval force, the Stork sloop of war being condemned and sunk and the Earl of Bathurst store ship disarmed, ready to be burnt on the approach of the enemy."[16] The two sloops were joined in May by H. M. S. *Mentor* (20) under Captain Robert Deans. Guard ships were then stationed inside the mouth of the harbor, and as frequently as the maintenance of wooden sailing ships allowed, one or another of the naval vessels patrolled the coast between Pensacola and Ship Island. Spanish prizes were taken upon occasion, but it was impossible to afford protection for British merchantmen desirous of sailing for England. Both *Mentor* and *Hound* took part in the abortive British strike against Mobile in January 1781. Dauphin Island was raided and a few prisoners taken, but the military force sent against the Spanish post on the east side of Mobile Bay was driven off. The next month, Deans decided to send the merchant vessels home, and on February 25 *Hound* sailed with a convoy of nine ships. A week out of Penascola they sighted Bernardo de Gálvez's invasion fleet. The convoy escaped, but on March 9 the Spaniards anchored off Santa Rosa Island.

Gálvez promptly landed troops on Siguenza Point, and heavy Spanish guns forced *Mentor* and *Port Royal* to withdraw from action. Undeterred by the misplaced Naval Redoubt atop the Red Cliffs, Gálvez led his ships into Pensacola harbor while British naval guns and crews were sent ashore to aid in the defense of the town. The story of the siege need not be retold. Gálvez proved himself to be a brave, competent commander, and he massed more that seven thousand men, including Spanish and French regulars, against General Campbell's motley collection of redcoats, Waldeckers, seamen, Indians, and volunteers. As Robert Deans put it, "The enemy counted thousands for our hundreds."[17]

Whatever his role in the Pensacola volunteers, George Gauld's skills were required by General Campbell during the siege. Writing with the naval surveyor's manuscript journal at hand, the compiler of *An Account of The Surveys of Florida* noted that Gauld "served as a volunteer during the siege, often assisted in the Engineer's department, and was of considerable service on many occasions. . . . From his

---

16. McNemara to Admiralty, August 6, 1780: ADM 1/2122.

17. Robert R. Rea and James A. Servies, *The Log of H.M.S. Mentor 1780–1781: A New Account of the British Navy at Pensacola* (Gainesville, 1982); Starr, *Tories, Dons, and Rebels*, pp. 193–213; Deans to Admiralty, May 20, 1781: ADM 1/1709.

knowledge of both the French and Spanish languages, he [was] employed by the General in Translating and interpreting all the letters and papers relative to Public Affairs, &c." When it was all over, Campbell acknowledged that "Mr. Gauld . . . deservedly merits my particular thanks for his Zeal and good Conduct in the lines during the siege of Pensacola."[18]

The explosion that destroyed the Queen's Redoubt on Gage Hill on May 8 extinguished the last British hopes. Within hours Campbell raised the white flag, and on May 10 the formal surrender of Pensacola was accomplished. The articles of capitulation provided for the removal of the military forces and civil officials of West Florida to New York, and with ample Spanish shipping at hand Gálvez rid himself of most of the British population in less than a month. Transports flying the flag of truce loaded during the last days of May and were dispatched, beginning June 4, for Havana.

Mr. and Mrs. George Gauld probably embarked at Pensacola about June 1 and reached New York by mid-July, together with their old friend Dr. John Lorimer. Gauld carried with him the maps and charts on which he had labored during the past sixteen years but little else. His property would be disposed of as best his colleague and agent John Mitchell could arrange. On August 10, 1781, Mitchell sold Gauld's Pensacola house and lot to Francisco Riaño of New Orleans for $600 cash. Under the circumstances, the surveyor was probably among the more fortunate exiles from West Florida.[19]

Gauld remained in New York for some time, and on August 18 he secured from General Campbell an introductory letter to Lord George Germain. Campbell praised Gauld's services and assured the American Secretary that as a result of his employment by the Admiralty, Gauld could "give your Lordship any Information you may be desirous of having."[20]

The surveyor might hope that there was yet a demand for his well-proven talents. When he was able, he made his way back to London. It was his intention "to solicit their lordships for some permanent es-

18. *An Account of The Surveys of Florida,* p. 3; Campbell to Germain, August 18, 1781: C.O. 5/597.

19. Another house and lot sold for a mere $90, but Gauld's Lot No. 175 was in a very desirable location. Riaño sold the property the next day to Antonio Lerin at the same price. Original sale documents in the Henry L. Huntington Library, San Marino, California; photostat copy, Board of Trustees of the Internal Improvement Trust Fund, Land Records Division (Tallahassee, Fla.).

20. Campbell to Germain, August 18, 1781: C.O. 5/597.

tablishment as he had all along been only paid by the day during the time he was employed,"[21] but he would have encountered long lines of suppliants at the Admiralty. The war was lost; in March 1782, Lord North's government collapsed. New faces filled the offices in Whitehall. Politics and peacemaking occupied their attention, and George Gauld had little time to wait. "From the nature of the climate and the fatigue of the services he had been so long employed upon, his health and constitution having been much impaired; in a few months after his arrival in London, he departed this life, and was interred in the burying ground of the Chapel in Tottenham Court Road."[22]

George Gauld died June 8, 1782. The site of his burial, Whitefield's chapel, suggests that within the hardy constitution that sailed so many leagues along the uncharted coasts of a distant colony lay a Christian spirit no less strong. His epitaph read, *"Mark the perfect man, and behold the upright: for the end of that man is peace."*

21. Jamaica, "Journal of the House of Assembly," vol. 8, 1784–91, November 18, 1789.
22. *An Account of The Surveys of Florida*, p. 27, identifies the final lines as *"Psalm XXX. verse 37"*; it is actually Psalm XXXVII.

CHAPTER XIII

Corpus Cartarum Vivarum

The life and work of the naval surveyor combined, in a unique fashion, both restless activity and disciplined application. His early voyages crossed the North Sea, then traversed the length of the Mediterranean. Twice he sailed the Atlantic; for a season he charted the waters of the Caribbean off Jamaica, and for twelve years he explored the coast, bays, and rivers of the Gulf of Mexico. Much of the time he worked in an open boat without protection from the elements. The blazing summer sun was his most constant companion and most welcome. Wind, waves, and chilling spray were unwelcome intruders that drove him back in search of the little sloop or schooner that offered scarcely more security or comfort. His comrades were the ordinary seamen bending at the oars, so mixed a lot as to defy description but sturdy tars if they survived the meager shipboard diet, diseases born of filth and tropical miasmas, and the dangers of the sea. For them, death was a commonplace to be put out of mind by the survey party's double ration of rum. The young lieutenant or midshipman placed in charge, a servant such as Peter Quash or John Robinson, or the rare professional partnership with John Payne provided the close associations of Gauld's working life.

Ashore his existence was quiet, almost solitary. His calling kept him at the drafting table, and the great charts demanded full attention if each line of coast and bar was to be perfectly inscribed. Such work was more exhausting, hour for hour, than the soundings and sightings that made it possible. If the risk was less, at Pensacola ink and paper were nearly as limited as rations were at sea. A torn page, a smudged line could border on catastrophe. No wonder that Gauld took such pains to seal, preserve, and guarantee delivery of his final drafts. His heart and soul went into them; they were a part of him.

The surveyor was an explorer in the truest sense and a scholar-scientist as well, living and working in two worlds with equal skill and dedication. He loved his work, but it was accomplished with a practical end in mind. His charts must make the sea safe for mariners, else the life of George Gauld went for naught. Therein lay his greatest disappointment, in all likelihood, but in contrast to so many of his fellow cartographers one hears no complaint. He was not a man to trumpet his virtues or peddle his own wares, and that seems to raise him above contemporaries whose fame surpassed his own. The quiet, self-effacing Scot was the Admiralty's surveyor first and last. To the navy he gave his hard-won knowledge and finely drafted charts. There is enough evidence from captains coming into Pensacola on "Gauld's marks," or in George Rodney's buoys guarding the shoals around Port Royal, to know that his record of discovery was also made available to the men who could best use it — and no way to know how many hand-drawn charts, made up for Royal Navy captains, fulfilled their purpose and were lost through wear and tear, salt spray, and time.

The cartographer's contribution could only be made permanent by printing, and that must wait upon the pleasure of Their Lordships at the Admiralty. Gauld was fully aware of the problem. In 1767 he sought Admiralty permission to publish his first perfected coastal chart, and in 1773 he raised the question of publication with Admiral Rodney. He must have hoped that his work at Port Royal would secure rapid dissemination, but that prospect disappeared when Rodney withdrew from the service after leaving Jamaica. In 1774 Gauld submitted certain "Remarks" on the Dry Tortugas to a Charleston newspaper, but his sailing instructions were not accompanied by an appropriate chart.[1]

The first of Gauld's charts to be printed, and the only one published during his lifetime, derived from his earliest surveying assignment at Pensacola. That initial project resulted in "A Plan of the Harbour of Pensacola in West Florida, Surveyed in the year 1764 by George Gauld M.A. The Bar by Sir John Lindsay." Succeeding years provided opportunities for the refinement and expansion of the manuscript, and the final revision was published August 1, 1780, by government command, in J. F. W. Des Barres's *Atlantic Neptune* series as *A Chart of the Bay and Harbour of Pensacola in the Province of West Florida.*[2] The Des Barres collection was worthy of Gauld's best, and inclusion of his chart should have been both a source of pride to him and a recommendation to other publishers. Unfortunately, it came at

---

1. Gauld to Rodney, November 8, 1773: ADM 1/239; *The South-Carolina and American General Gazette*, vol. 17, no. 825 (July 1–8, 1774), pp. 188–89.

2. Original in the Library of Congress. Lowery, *Descriptive List*, p. 398; P. Lee Phillips, *Geographic Atlases*, 3: 456. See also G. N. D. Evans, *Uncommon Obdurate: The Several Public Careers of J. F. W. Des Barres* (Peabody Museum, Salem, Mass., 1969), chap. 2.

a time when he was isolated in a distant, beleaguered colony; after he returned to England, he had little time to negotiate with prospective publishers.

It is likely that Gauld did seek Admiralty permission to publish his charts and discussed the problems of printing them when he was in London in 1782. He would naturally have turned to William Faden at No. 5, Charing Cross, for advice. Partner of Thomas Jefferys, the well-known distributor of maps and charts, until the older man's death in 1771, William Faden boasted the title of "Geographer to the King." His shop, maintained in the same location until his retirement in 1823, was the commercial center of British cartography, and its master enjoyed the highest reputation and honors. In 1795, Faden's work won a £50 prize and a gold medal from the Royal Society of Arts, and in 1796 a second gold medal was awarded him. He was much interested in the great mapping project known as the Ordnance Survey in the early nineteenth century and contributed significantly to the rapid growth of geographic knowledge and accurate mapping in Great Britain.[3] Not only did Faden eventually produce several of George Gauld's charts; he was also the recipient of a considerable body of Gauld's private papers, some of which he printed, others he referred to when describing the surveyor's career. The latter included a journal of the young naval schoolmaster's voyage to the Mediterranean in 1758, records kept during the surveying seasons in American waters, and a journal of the siege of Pensacola, none of which appear to have survived.[4]

Whatever the surveyor's personal relationship with Faden in 1782, he had already taken steps to ensure that his work would not be lost. On October 20, 1779, as the Spanish war began to engulf West Florida, George Gauld drew up his will, and in it he recognized that his charts would stand as lasting testaments to his life's work. After providing that any debts should be discharged, he designated gifts of mourning to his friends Dr. and Mrs. John Lorimer; to his sister Isabel Gauld and his uncle Farquhar Shaw, both of Botriphnie, he bequeathed £100 which had been deposited with Dr. George Hay of Keith. The residue of his estate he left to his wife, Ann, for they were childless. Ann was, of course, the legal executrix, but John Lorimer was appropriately named as Gauld's professional executor. To Lori-

---

3. Ida Darlington and James Howgego, *Printed Maps of London circa 1553–1850* (London, 1964), pp. 30–31.

4. *An Account of The Surveys of Florida*, pp. 3, 5, 25–27.

mer he gave his most valued instruments: a horizontal stopwatch, a telescope, and — most intriguing of all his possessions — his electrical machine. Lorimer was to have whatever books and manuscripts he might desire and was authorized to destroy all letters and papers he might think fit. More significantly, Gauld directed that if the Admiralty should be pleased to permit his charts to be published, that work should be pursued under Lorimer's supervision. For his trouble the surgeon should receive half of any profits that might accrue, the other half to go to Ann Gauld.[5] It is unclear whether the exact terms of the surveyor's will were fullfilled. Dr. Lorimer remained at New York in 1782-83, only leaving with the last British troops. He eventually settled in London, where he died in 1795.[6] He was, however, instrumental in securing the publication of Gauld's charts and acted as intermediary between Ann Gauld and publisher William Faden.

By her own description, Mrs. Gauld was left "a helpless widow" when her husband died. Her sad plight need not be doubted in view of the circumstances of most West Florida exiles in London and the cruel if correct attitude of the British government toward them. Gauld had enjoyed a respectable salary and doubtless lived in a frugal manner, but any reserve he might have accumulated was wiped out by his London agent's bankruptcy in 1776, and the return from the sale of his property in Pensacola must soon have been expended. His salary was paid to the date of his death, but nearly a year passed before his widow could collect the final installment.[7] Aware of her late husband's hopes, Ann Gauld must have sought the advice of both John Lorimer and William Faden in order to put Gauld's material to practical and profitable use.

William Faden was a shrewd businessman as well as a prestigious publisher. He would have insisted that Ann Gauld secure permission from the Admiralty for the reproduction of the surveyor's charts, and he would have suggested that a subsidy for publication was most de-

5. PROB 11/1091. In addition to Ann Gauld and John Lorimer, John Mitchell was also named an executor. The will was witnessed by Edmund Rush Wegg, John Falconer, and Alexander Moore, all prominent Pensacolans. It was proved at law, June 22, 1782, in the Prerogative Court of Canterbury by oath of Ann Gauld, who was made sole executrix.

Examples of contemporary "electrical machines" may be seen in Wheatland and Carson, *The Apparatus of Science*, pp. 136-40. They generated static electricty by spinning a glass globe against a leather pad.

6. Robert R. Rea and Jack D. L. Holmes, "Dr. John Lorimer and the Natural Sciences in British West Florida," *Southern Humanities Review* 4 (1970): 370, 372n.25.

7. Admiralty Board Minutes, March 12, 1783: ADM 3/97; see also ADM 12/55.

sirable, perhaps even necessary. Dr. John Lorimer applied to the Admiralty in Ann Gauld's behalf, and on April 1, 1788, Their Lordships granted permission for her "to publish at her own risque, the Chart made by her late Husband containing the Tortugas, Martyrs, & Florida Keys." As the cost of engraving and printing was beyond her means, Mrs. Gauld turned to the Jamaica House of Assembly for financial support. Her petition, presented November 18, 1789, set forth George Gauld's services, his "assiduity and attention . . . for upwards of seventeen years," as a result of which his "health and constitution [were] much impaired," and sought the assembly's assistance toward the publication of those parts of her "late husband's surveys, which particularly concern the island of Jamaica." The petition was "referred to the committee of the whole house to inquire into and take further into consideration the state of the island." Out of their deliberations came a resolution, first read on November 25 and adopted one week later, which "recommended to the house to direct the agent to pay the sum of one hundred guineas to the order of Mrs. Ann Gauld, widow of George Gauld, surveyor, deceased, whenever the surveys of the Gulph Passage shall be properly published, and approved of by men of skill in that profession; an accurate survey of that passage being much wanted."[8]

The generosity of the Jamaica Assembly should not be overrated, for its motives were by no means disinterested. One hundred guineas was probably little more than enough to meet the initial expense of engraving and publishing Gauld's charts, and the members of Jamaica's Assembly had an immediate interest, as merchants and planters, in the safe passage of their cargoes to England. William Faden pointed out another group who stood to gain significantly if "Captains and Masters of vessels . . . make use of those charts." A publisher's footnote in *An Account of The Surveys of Florida, &c. with Directions for Sailing . . . through the Gulph of Florida* observes that "The Underwriters will undoubtedly profit more by this Publication, than any other set of men whatever; provided they take care, that every West Indies vessel they insure, shall have a set of Mr. Gauld's charts on board."[9]

The Jamaican subsidy provides a remarkable example of transoce-

8. Admiralty Board Minutes, April 1, 1788: ADM 3/104; Jamaica, "Journal of the House Assembly," vol. 8, 1784-91, November 18, 25, December 2, 1789, West Indian Reference Library, Institute of Jamaica, Kingston.
9. Page 21.

anic interest in Gauld's work. One can only wonder at the personal and political connections that brought about the assembly's action and hope that its intentions were indeed fulfilled.[10] The prospects were sufficiently promising to gain the attention of the *New York Daily Advertiser,* which noted on March 4, 1790: "We congratulate our nautical friends of its being proposed to publish the accurate surveys of the gulph passage made by the late Mr. George Gauld of West Florida." The announcement further reported the measures taken by the Jamaica Assembly on November 25 "to encourage the publication."[11] Commercial activity was international, and Gauld's charts were of as much interest to Yankee skippers as to British.

It is obvious that William Faden was encouraged to invest in the production of Gauld's work by Ann Gauld's successful plea and the promise of a hundred guineas. In April and May 1790 he brought out a series of Gauld's charts: *The Grand Caymans,* priced at one shilling, *The West End of the Island of Cuba and part of the Colorados,* and *An Accurate Chart of the Tortugas and Florida Kays or Martyrs,* the last offered at twelve shillings. All were published with Admiralty permission.[12] In conjunction with these detailed charts Faden published *An Account of The Surveys of Florida, &c. with Directions for Sailing from Jamaica or the West Indies, by the West End of Cuba, and through the Gulph of Florida,* a twenty-seven-page pamphlet which included "A General Chart to accompany Mr. Gauld's Surveys of the Florida Keys &c." Compiled by Faden from the notes on Gauld's manuscript charts, the *Account* provided a handy companion volume to the large navigational charts. "A Concise Account of Florida for the Space of Two Hundred and Seventy Years" (1512–1782) served as Introduction and included a brief "extract of Mr. Gauld's journal in manuscript" relating to the siege of Pensacola. Detailed sailing directions made up the core of the volume and were supplemented by remarks upon watering places, winds, and currents. Interspersed among these sections was a priceless body of biographical and historical information relative to George Gauld, which demonstrates that Faden had some part, if not all, of

10. The colonial agent, Stephen Fuller, was resident in England at this time, but his correspondence in the Jamaican Archives is incomplete and the period October 21, 1789, to July 12, 1792, is missing (correspondence between John Ware and Jacqueline Welds, September 2, October 13, 1971, and Clinton V. Black, archivist, October 26, 1971).

11. *New York Daily Advertiser,* vol. 6, no. 1572, March 4, 1790.

12. British Museum, *Catalogue of Printed Maps, Charts and Plans* (London, 1967), 6: 534; L. S. Dawson, *Memoirs of Hydrography* (1883–85; reprint ed., London, 1969), pt. 1, p. 4. A second edition of the last-named chart appeared in 1820.

the surveyor's private papers at his disposal. Unhappily, he seems to have been the last person to make use of those manuscripts; they are no longer to be found.

The *Account of The Surveys of Florida* also seems to owe something to Dr. John Lorimer. Several references to earlier scientific essays by Lorimer and to his interest in terrestrial measurement appear as footnotes in the pamphlet. Furthermore, among the subscribers to the first of George Gauld's posthumous publications was His Royal Highness the Duke of Clarence, with whom Dr. Lorimer was acquainted. Gauld's old friend had encountered the future King William IV during the closing days of the American Revolution, when he was inspector general of military hospitals at New York and "Sailor Billy" was a "very ingenious young Midshipman" who visited that harbor. Prince and physician met, and Lorimer prepared for the royal apprentice a brief essay on magnetism which he later expanded into a book dedicated to the duke and published by none other than William Faden.[13] Whatever his role, John Lorimer would have rejoiced to see Gauld's work in print.

Faden was proud of his coup, as well he might be. These were the British Admiralty's specially ordered surveys, the finest charts of the Gulf passage extant, "faithfully copied from the Originals in the Admiralty Office." As the "General Chart" or "Litte Chart" covered a larger area than Gauld had actually surveyed, Faden cited the additional authorities he had used in putting it together: De Brahm for part of the Florida peninsula and Don Juan Lopez for Cuba — although Faden warned that "Lopez's map of Cuba, is nothing more than what was published by Mr. Jefferys in 1762, *Hispanized.*"[14]

The publisher also disclosed a hitherto unnoticed example of the expropriation of Gauld's work by another author. Faden declared that the surveyor's chart of Espíritu Santo, made in 1765, was published in Dr. William Stork's *A Description of East-Florida* (3d ed., 1769). "How it happened to be published there we cannot pretend to say; but it is the only part of the many surreptitious sketches, which

---

13. Rea and Holmes, "Dr. John Lorimer," *Southern Humanities Review* 4 (1970): 369–70; *An Account of The Surveys of Florida*, pp. 6, 21–22; Gauld, *Observations*, p. 22. These references might have been incorporated by Gauld had he prepared his manuscript for publication in 1781 or 1782, but that seems unlikely. Perhaps Dr. Lorimer acted as amanuensis for his deceased friend; perhaps Faden was simply puffing a forthcoming book. Prince William's logbooks shed no light on his contact with Lorimer: Log N/B/8–9 (NMM).

14. *An Account of The Surveys of Florida*, pp. 7–8.

have been pirated from Mr. Gauld's works, that has been literally and pretty correctly copied."[15] This statement is fascinating. Stork's original publication in 1766, entitled *An Account of East-Florida*, made no use of Gauld material, but the 1769 "Third Edition" of his pamphlet contained two maps that reveal unmistakable evidence of Gauld's cartography. Stork's "Bay of Espíritu Santo" is nothing less than an engraved version of Gauld's unpublished 1765 chart "A Survey of the Bay of Espíritu Santo in East Florida," and Stork's inclusive chart of "East Florida" repeats and confirms the plagiarism of Gauld's Tampa Bay survey. More intriguing is the fact that William Faden should call attention to Stork's thievery and that he should protest his innocence of the means by which it occurred; for Dr. Stork's maps were "Sculp'd" and sold by Thomas Jefferys, who was Faden's senior partner in 1769, and in 1774, three years after Jeffery's death, the firm, then "Faden and Jefferys," brought out the suspect maps in a new version of Stork's pamphlet.[16] It would appear that Faden was trying to cover his tracks—and did so clumsily.

The publication of charts and sailing instructions was vital to mariners but not likely to produce immediate profits for printers and booksellers. In addition to Ann Gauld's subsidy from the Jamaica Assembly, William Faden sought other sources of support. Declaring that he gave Gauld's charts to the public in order to prevent maritime accidents, he added, "We *say given,* for the price that has been put upon this necessary work, never could have defrayed the expense thereof." Scholarly or scientific publication usually depended upon the support of a numerous or notable body of subscribers in the eighteenth century. In the case of the "Florida gulph" charts and commentary, Faden acknowledged that he was "generously aided by [the] subscription" of "the Lords Commissioners of the Admiralty, His Royal Highness the Duke of Clarence; and several Gentlemen of the Island of Jamaica, who knew the value of these charts."[17] It was a potent combination, successful in bringing an important part of George Gauld's work into print and, it may be hoped, providing Ann Gauld some financial security.

15. Ibid., p. 6.
16. William Stork, *A Description of East-Florida* (London, 1769), plate, and p. 35; compared with Gauld, "A Survey of the Bay of Espiritu Santo in East Florida . . . 1765," NLMD, no. 41. Numerous versions of Stork's work were published between 1766 and 1769, most by the firm of W. Nicoll and/or G. Woodfall.
17. *An Account of The Surveys of Florida*, p. 20.

G^eo. Gauld — Cartographer

No better evidence of a demand for the surveyor's productions could be found than Faden's publication of further examples in succeeding years. In 1794 there appeared *A Chart of the Gulf of Florida or New Bahama Channel, commonly called the Gulf Passage, between Florida, the Isle of Cuba, & the Bahama Islands: From the Journals, Observations, and Draughts of Mr. Chas. Roberts, Master in the R[oyal] Navy, Compared with the Surveys of Mr. George Gauld &ca.* which utilized certain parts of Gauld's work.[18] Then, in 1795, Faden produced Gauld's *Remarks and Directions concerning the Channels and for Sailing into Port Royal and Kingston Harbours; with cautions, to avoid the dangers that lie in the way*, whose fourteen-page text was taken directly from the notes on the 1772 manuscript "Plan" of Port Royal.[19] The chart itself was published January 1, 1798, as *A General Plan of the Harbours, of Port Royal and Kingston Jamaica. With the Channels leading thereto, and the Kays and Shoals adjacent Including Wreck Reef.* The "Remarks" are properly attributed to "George Gauld, M.A. Surveyor General of the Coast of West Florida, &c.", but the chart is not credited to him, strangely enough. Although Faden used Gauld's title verbatim and patently drew upon his manuscript chart, he made some changes — including the elimination of all reference to Admiral Sir George Brydges Rodney, even the identification of his house outside Kingston. A new edition of Gauld's *Remarks and Directions* was issued at the same time, although Faden did not identify it as a reprinting of his earlier publication.[20]

In 1796, Faden had published Gauld's *Observations on the Florida Kays, Reef and Gulf; with Directions for Sailing Along the Kays, From Jamaica by the Grand Cayman and the West End of Cuba: also, A Description, with Sailing Instructions, of the Coast of West Florida, Between the Bay of Spiritu Santo and Cape Sable.* This twenty-eight-page volume reproduced the technical content of the earlier *Account of The Surveys of*

---

18. Lowery, *Descriptive List*, pp. 339, 429. These surveys also formed the basis for the I. W. P. Lewis *Chart of the Dry Tortugas and part of the Florida Reef shewing the channels to Key West harbour and the adjacent islands compiled from the Surveys of George Gauld, esq.* (Boston: Benjamin Loring & Co., 1838).

19. An interesting indication of Gauld's care in updating his charts appears in the note stating that John Payne had discovered another bank near the Harbour Shoal when in the *Thynne* packet, April 10, 1774.

20. Compare MODHD, A640 Ag 4 and the printed chart in the British Museum Map Room. *Remarks and Directions* . . . Entered at Stationers Hall. London: Printed for W. Faden, 1798, VII/54, British Museum Map Room.

Florida with certain corrections of latitude and longitude, minor rearrangement of material, and some stylistic revisions. In order to make the *Observations* more useful, Faden included a "Description of the East Shore of Florida, by M. Gerard de Brahm," and Gauld's "Instructions for the West Shore of Florida between the Bay of Spiritu Santo and Cape Sable." Finally, the book included some "General Directions for Sailing from Cape Canaveral to Kay Biscaino, and from thence, within the Florida Reef, into the Gulf of Mexico. By Captain Barton of Carolina. Communicated to Mr. Gauld, October 24th, 1774."

Evidently the publication of Gauld's *Account* in 1790 had attracted the interest of seafarers, for their comments and corrections were reflected in the *Observations*. The later volume is more efficiently organized, printed in a bolder type, and would be more useful as a mariner's handbook. Its publication might suggest that the *Account* had sold out; at the least, Faden's editorial efforts lead to the inescapable conclusion that a new book by George Gauld was apt to be welcomed by the maritime fraternity.

The next of Gauld's works published by William Faden was printed in 1803: the monumental *An Accurate Chart of the Coast of West Florida, and the Coast of Louisiana; from Sawaney River on the West coast of East Florida to 94°20' West Longitude*. The most extensive of the Gauld charts and the most comprehensive of his many surveys, it covered some 850 nautical miles of the northern and eastern coast of the Gulf of Mexico, delineated the many harbors, bays, sounds, and rivers, recorded depths and bottom characteristics ranging beyond the hundred-fathom curve, and extended inland many miles at certain points. The title of the chart indicated that the individual surveys upon which it was based were made between 1764 and 1771, but the chart also incorporated the discoveries of Gauld's last voyage to the west in 1777, when he cruised the coast of Spanish Louisiana. The task of bringing together all these various drafts and surveys and reducing them to a common scale was effectively completed by Gauld in 1778. It was probably his last cartographic project, the final strokes having been added early in 1779 as he waited for Captain Joseph Nunn's *Hound* to be refitted after the hurricane of October 1778, in order to deliver it safely to Sir Peter Parker. This magnificent chart was engraved for Faden by B. Baker of Islington, a master of his craft, and was printed on four large sheets which sold for sixteen shillings.[21] It is a triumph of chartmaking, an appropriate culmination to the life work of George Gauld. Appropriately, the chart was republished twenty years later by the British Hydrographical Office as Admiralty Chart No. 524.[22] Thus the purpose for which Gauld was sent to the inhospitable shores of West Florida and the unknown waters of the Gulf of Mexico was finally achieved.

Faden made further use of Gauld's surveys in *A General Chart of the West Indies and Gulf of Mexico, describing the Gulf and Windward Passages Coasts of Florida, Louisiana and Mexico.* Published June 4, 1808, this composite chart was "Drawn from the Surveys taken by Mr. Geo. Gauld, and others, the New Spanish Charts, &c. and adjusted from recent Observations, by P. Foss Dessiou, Master of the Royal Navy" and carried the approval of the Chart Committee of the Admiralty.[23] Seven years later the navy's continuing appreciation of Gauld's work was demonstrated by the republication of *The Island of Grand Cayman*, originally surveyed in 1773.[24] After more than forty years, George Gauld's survey was still the best available. In fact, in 1876, when the Admiralty published a new chart of "North America—East Coast, Florida Reefs: Boca Grande Cay to Tortugas Cays," which incorporated both British and U.S. Coast Survey findings of 1873 and 1874, the Marquesas Cays and certain extended soundings were taken "from Gauld's survey 1775."

International recognition of Gauld's accomplishments came first, ironically, from Spanish cartographers who adapted his published charts for their own use. The undated *Carta Esférica* was a Hispanicized version, similar in virtually every detail, of Gauld's *Chart of the Gulf of Florida*. The title (translated) acknowledged that it was "the survey made by George Gauld by order of their Lordships of the British Admiralty in the years 1773, 74, 75, of the Florida Keys from the cited survey of the Tortugas to the southwesternmost [part] of Key Largo." Gauld's longitudes were said to have been "corrected . . . by the observations of Brigadier of the Royal Navy Don Thomas

21. Dawson, *Memoirs*, pt. 1, p. 4. Dawson credits Gauld with a survey of "St. John's Harbour," but no record of such a survey appears.

22. Much of the work reflected in the two versions of this chart found its way into a somewhat similar edition published by Edmund and G. W. Blunt in 1828 and 1830. See Lowery, *Descriptive List*, p. 338; Phillips, *Geographical Atlases*, 3: 478.

23. F.O. 925/4340 (P.R.O.).

24. Published according to Act of Parliament by Capt. Hurd RN Hydrographer to the Admiralty 1st. Decr. 1815. Sec. VIII (462) NS 9, British Museum Map Room.

Ugarte, *Teniente de Navío* Don Mariano Ysasbiribil," and his soundings converted to "fathoms of six feet Burgos." A comparison of the coordinates of Gauld's *An Accurate Chart of the Tortugas and Florida Kays or Martyrs* and the *Carta Esférica* discloses that the "corrections" made by the distinguished Spanish naval officers consisted of nothing more than relating Gauld's coordinates, based on Greenwich, to another prime meridian, probably that of Cadiz.[25]

Eventually the great *An Accurate Chart of the Coast of West Florida, and the Coast of Louisiana* (1803) was also given a Spanish accent and found its way into, of all places, the archives of the Spanish army. It appears in exactly the same form as that published by Faden, including the English title and English toponymy or list of place-names on the four sheets. The provenance of this map is unknown, but obviously it postdates 1803 and is unlikely to have been made later than 1821, the year the Floridas were finally acquired by the United States.[26]

The British coastal surveys were, of course, known to the succeeding generation of American cartographers. Andrew Ellicot remarked that:

Mr. Gauld's survey of the Dry Tortugas and the Florida reef and Keys . . . may justly be considered as one of the most valuable works of the kind extant. . . . From Key Vaccas to Key Largo, I carefully compared Mr. Gauld's charts with the soundings, and perspective views of the Keys, and found an agreement which excited my surprise, and am induced to believe that not a single rock or shoal, so far as the work extends, has been omitted, and that not an error of three feet will be found in any of the soundings. If this work had been completed, it might be esteemed one of the most perfect and useful of the kind.

Ellicott was acutely aware that "a knowledge of this navigation is of very great importance to the mercantile interest of the United States," and when he completed his own survey, he deposited his copy of Gauld's chart in the office of the secretary of the navy.[27]

---

25. *Carta Esférica*, Museo Naval, Madrid.

26. Republished by the Servicio Geográfico del Ejército, Año 1803, *Numero* 106, "An Accurate Chart of the Coast of West Florida, and the Coast of Louisiana," in four sheets (88 x 66 cm.), Classification P-b-2-13, in *Cartografia de Ultramar, Carpeta II, Estados Unidos y Canada* (Madrid, 1953), pp. 402–12.

27. Andrew Ellicott, *The Journal of Andrew Ellicott* (1803; reprint ed., Chicago, 1962), pp. 255–56, 273.

G^eo. Gauld — Cartographer

Nearly half a century would pass, however, before the United States government recognized the desperate need to carry on the work that Gauld had begun.

In 1849, four years after Florida achieved statehood, Alexander Dallas Bache, superintendent of the U.S. Coast Survey, reported to the secretary of the treasury that during the previous year forty-one vessels had been wrecked on the Florida Keys or had put into Key West in distress. Of these, seventeen were total losses or condemned. To further illustrate the dangers of the passage into the Gulf of Mexico, Bache cited the high insurance rates demanded for vessels in this trade, rates which were comparable to those for vessels plying the most hazardous routes. "The charts of this part of our coast are," he wrote, "with the exception of detached surveys of some of the harbors, and of portions of the Florida reef, inaccurate and wanting in details. Gauld's chart, which is one of the most authentic of any extended portion of the coast, dates as far back as 1790."[28] Bache's reference to the distant publication date of the chart did not, of course, take into consideration the fact that it was actually based on Gauld's surveys during the years 1773–75.

Included in the superintendent's report was a letter from Lieutenant C. R. P. Rodgers, U.S. Navy, whose praise for Gauld's work was based upon personal experience. He wrote in part:

During the last three years of the Seminole war, I served in the flotilla cooperating with the army, and was actively employed in cruising among the Florida reefs and "keys." I found the charts of the peninsula very inaccurate, with the single exception of Gauld's chart of the Florida reef, published in 1790, which was evidently a work of merit, and is still valuable, although important changes have taken place since its publication.

Lieutenant Rodgers's knowledge and appreciation of Gauld's chart did not extend to his other posthumous publications, for the American believed that this survey "was ended by the death of Gauld, who, it is said, was buried on the spot where his labors terminated."[29]

---

28. Bache to Secretary to the Treasury, February 15, 1849: *Report to the Secretary of the Treasury communicating a report from the Superintendent of the Coast Survey, in relation to the survey of the coast of Florida*, Exec. Doc. No. 30, 30th Cong., 2d sess., p. 2. On the coast survey see A. Hunter Dupree, *Science in the Federal Government* (New York, 1957), pp. 29–33, 52–56, 100–105.

29. Rodgers to Bache, December 28, 1848: ibid., p. 8.

Stephen R. Mallory, then collector of customs at Key West, also advised Superintendent Bache that

The portion of these reefs which has proved most destructive to commerce, is that which lies between Indian Key and Key Biscayne, a distance of about eighty miles. No American survey has ever been made of it, and that of Gauld, if I am not mistaken, embraced only a part of Carysfort Reef . . . Gauld's survey with Blunt's compilation of it and a few old Spanish charts furnish the sum of our information upon this subject and no mariner of ordinary prudence would attempt to cross the Florida reef with these only as his guide.[30]

Although Mallory's letter demonstrates a thorough knowledge of shipping along the Florida Keys and the hazards of the route, he clearly knew less about Gauld and his work than did Bache and Rodgers; but with such encouragement and support the task of the U.S. Coast Survey and its successors would be carried on toward completion.

As for George Gauld, naval surveyor, the work survived the man, as it should, highly prized by seafarers of three nations whom fate momentarily conjoined while he traced the Gulf Coast outline of North America. Britons, Spaniards, and Americans came to build, and so changed the face of the land that the eighteenth-century heritage all but disappeared. Cities rose in the wilderness. Where Gauld once marked the sentinel tree along the dunes, towers of steel and glass now stand. Even the contours of bays, channels, and anchorages have changed, for nature as well as man has taken a toll of George Gauld's coast. Only the sea remains unchanged, for it is ever changing. That — and the spirit of men like Gauld who work to make safe its ways for other men — remains.

30. Mallory to Bache, December 28, 1848: ibid., pp. 9–13. The reference is to E. and G. W. Blunt, *Blunt's charts of the north and south Atlantic oceans, the coast of North and South America, and the West Indies* (New York, 1830). See Phillips, *Geographical Atlases*, 3: 478.

APPENDIX

A Note on the Sources and a Checklist of the Works of George Gauld

THE SOURCES

The basic manuscript sources bearing upon the life of George Gauld deserve a brief comment, and his own works require bibliographical attention, for scattered as they are, no extant listing encompasses the whole of his surviving charts, descriptions, published maps, and sailing instructions. The contributions of secondary authorities are duly noted throughout the text.

The greater part of the pertinent manuscripts is to be found among the Admiralty papers located in the Public Record Office in Kew. Of primary value are the logs of the ships on which Gauld served and of ships that were in the same waters at the same time. In almost every case, both the captains' logs (ADM 51) and the masters' logs (ADM 52) were consulted, for in spite of their common source — the masters' daily accounts — they frequently provide different and complementary information. The journals kept by lieutenants aboard these vessels (L/—) were also used, as far as they survive, at the National Maritime Museum, Greenwich. Only in the case of the two surveying ships *Florida*, whose logbooks have disappeared, did this source fail to provide a continuous record of Gauld's movements at sea.

A second important body of documents is the correspondence of the admirals on the Jamaica Station and the captains under whom Gauld sailed (ADM 1). Gauld reported to Port Royal on a fairly regular schedule, and the admirals abstracted or forwarded his letters to London. One of the surveyor's autograph letters is filed under his name among the captains' letters. Unfortunately, the observations of the lieutenants who commanded the little surveying sloops have not survived, save as they were occasionally forwarded from Jamaica to London or incorporated in an admiral's report.

A third useful class of documents consists of the muster tables of the ships to which Gauld was attached (ADM 36). As the surveyor

235

Appendix

was often absent from these ships, the value of this material is limited; but in the case of the sloop *Florida*, the muster provides the key to establishing the chronology of Gauld's work from 1774 to 1777.

Other Admiralty records, and the Navy Board correspondence at Greenwich, contain occasional references to Gauld and are cited where pertinent. The Colonial Office papers (C.O. 5) bearing upon West Florida, also at the P.R.O., provide most of the information regarding Gauld's life ashore.

Special notice must be given to the unrivaled collection of Gauld's manuscript charts held by the Ministry of Defence Hydrographic Department, at Taunton, and listed in *Hydrographic Department Professional Paper No. 13*, "A Summary of Selected Manuscript Documents of Historical Importance Preserved in the Archives of the Department." The charts at Taunton are the mother lode from which the cartographic content of this study was extracted.

CHECKLIST OF THE WORKS OF GEORGE GAULD

Manuscript Charts

The Florida Keys

A Plan of the Tortugas and Part of the Florida Kays; Surveyed by George Gauld M.A. 1773. For the Right Honourable the Board of Admiralty. [MODHD, D966 88.]

Rough field sheet, Dry Tortugas and Florida Keys. [MODHD, U9 6g.]

A Plan of Part of the Florida Kays, from Bahia Honda to Cayo Largo; Surveyed by George Gauld M.A. For the Right Honourable the Board of Admiralty. 1775. [MODHD, D959 88.]

Rough draft, the Florida Keys from Rodriguez Key to the Dry Tortugas. [MODHD, D959a 88.]

Espiritu Santo

A Survey of the Bay of Espiritu Santo in East Florida By Geo: Gauld M.A. 1765. [MODHD, x64 Jv.]

A Survey of the Bay of Espiritu Santo in East Florida By Geo: Gauld M.A. in the year of our Lord 1765. [Guard Book Series of Manuscript Maps, No. 41, MOD Naval Library, London.]

A Survey of the Bay of Espiritu Santo in East Florida By Geo: Gauld M.A. By Order of Sir William Burnaby Rear Admiral of the

Red &c. &c. 1765. [Guard Book Series of Manuscript Maps, No. 42, MOD Naval Library, London.]

The Coast of East and West Florida

A Survey of the Coast of West Florida from Pensacola to Cape Blaise: including the Bays of Pensacola, Santa Rosa, St. Andrew, and St. Joseph, with the Shoals lying off Cape Blaise. By George Gauld M.A. For the Right Honourable the Board of Admiralty. 1766. [MODHD, A9464 31c.]

Rough field sheet, Sawaney River to Cape Blaise. [MODHD, E 19/12 4a.]

Rough field sheet, Cape Blaise to Perdido Bay. [MODHD, E 19/9 5e.]

A Sketch of the Entrance from the Sea to Apalachy and Part of the Environs taken by George Gauld Esqr. Surveyor of the Coast and Lieutenant Philip Pittman Asst Engineer. [William L. Clements Library, Ann Arbor.]

A Plan of the Harbour of Pensacola in West-Florida Surveyed in the Year 1764 by George Gauld M.A. The Bar by Sir John Lindsay. [G. & M. Div., Group RMR, Library of Congress.]

A Survey of the Bay of Pensacola with part of Sta. Rosa Island &ca. Geo: Gauld fecit. 1766. [Inset "Plan of Pensacola" and "Plan of Campbell-Town." C.O. 700, Florida No. 32, P.R.O.; Guard Book No. 47, MOD Naval Library.]

Survey of Pensacola and its Environs, by George Gauld, Lieut. Durnford and other British Engineers, copied from the original in possession of Wm. Tatham by T. Stephenson March 1813. [Record Group 77, L44, Office of the Chief of Engineers, Cartographic section, National Archives, Washington, D.C.]

A Plan of the Bays of Pensacola and Mobile with the Sea Coast and Country adjacent By Geo: Gauld M.A. For the Right Honble. the Board of Admiralty 1768. [MODHD, D964 Rt.]

Rough field chart of Mobile Bay. [MODHD, E 19/10 5k.]

Rough field chart of Mississippi Sound, the Chandeleurs to Lake Pontchartrain. [MODHD, D962 88.]

Rough field chart, Dauphin Island to Lake Maurepas. [MODHD, E 19/13 4a.]

Appendix

The Mississippi River

A Plan of the Mouths of the Mississippi By George Gauld M.A. ["For the Right Honourable the Earl of Dartmouth." M.P. G530, P.R.O.]

A Plan of Manchac 1774. [Inset "Part of the Mississippi near Manchac 1774." "For His Excellency Sir Henry Clinton K.B. Commander in Chief &c. &c. &c. at the request of Brigr. General Campbell 1779." Clinton Papers, William L. Clements Library, Ann Arbor.]

A Plan of the Coast of Part of West Florida & Louisiana including the River Mississippi from its Entrances as high up as the River Yazous. Surveyed by George Gauld M.A. for the Right Honourable the Board of Admiralty. ["This survey has been taken at different times, and reduced to one general scale in the year 1778." Inset "A Plan of Manchac 1774." MODHD, D958 88.]

Spanish Louisiana

A Draught of Part of the Coast to the Westward of the River Mississippi with Part of the Island of New Orleans &c. By George Gauld M.A. For the Right Honourable the Board of Admiralty 1777. [MODHD, D965 88.]

Rough field sheet, mouths of the Mississippi to Galveston Bay. [MODHD, E 19/14 4a.]

Port Royal, Jamaica

A General Plan of the Harbours of Port Royal and Kingston Jamaica, with the Channels leading thereto, and the dangerous Shoals adjacent, including Wreck Reef &c. For the Right Honble. the Board of Admiralty. By George Gauld 1772. [MODHD, D961 16n.]

A General Plan of the Harbours of Port Royal and Kingston Jamaica with the Channels leading thereto, and the Kays & Shoals adjacent; including Wreck Reef &c. By George Gauld 1772. For Sir George Brydges Rodney Bart. Vice Adml. of the Red, & Commr. in Chief at Jamaica &c. &c. [MODHD, A640 Ag 4.]

Rough draft. [MODHD, B216 Ag 2.]
Rough draft. [MODHD, r26 Ag 1.]

G^eo. Gauld — Cartographer

Grand Cayman

The Island of Grand Cayman. Geo. Gauld fecit 1773. For the Right Honble. the Board of Admiralty. [MODHD, q43 Ag 1.]

The Island of Grand Cayman Surveyed by Geo. Gauld 1773. [MODHD, 196 Ag 1.]

The Island of Grand Caymana. [MODHD, U 10 Ag 1.]

Published Charts

A Chart of the Bay and Harbour of Pensacola in the Province of West Florida Surveyed by George Gauld A.M. Published by Command of Government by J. F. W. Des Barres Esqr. Aug. 1, 1780, in *The Atlantic Neptune.*

An Accurate Chart of the Tortugas and Florida Kays or Martyrs, Surveyed by George Gauld, A.M. in the Years 1773, 4, & 5. By Order of the Right Honourable the Lords Commissioners of the Admiralty, And now Published by permission of Their Lordships. London: Wm. Faden, April 5, 1790; 2d ed., 1820.

The West End of the Island of Cuba, and part of the Colorados. Surveyed by Geo: Gauld A.M. in 1773 [published together with] *The Grand Caymans Surveyed by Geo. Gauld A.M. 1773.* London: W. Faden, May 12, 1790; 2d ed., 1820.

The Island of Grand Cayman, Surveyed by George Gauld 1773. Published according to Act of Parliament by Capt. Hurd RN Hydrographer to the Admiralty, Dec. 1, 1815.

A General Plan of the Harbours, of Port Royal and Kingston Jamaica. with the Channels leading thereto, and the Kays and Shoals adjacent Including Wreck Reef. By George Gauld, M.A. Surveyor General of the Coast of West Florida, &c. London: W. Faden, Jan. 1, 1798.

An Accurate Chart of the Coast of West Florida, and the Coast of Louisiana; from Sawaney River on the West Coast of East Florida to 94° 20' West Longitude Describing the Entrance of the River Mississippi, Bay of Mobile, Pensacola Harbour &c. Surveyed in the Years 1764, 5,6,7,8,9,70 & 71. By George Gauld M.A. under the directions of the Right Honourable the Lords Commissioners of the Admiralty. London: W. Faden, Feb. 4, 1803, 2d ed., 1820; published by the Hydrographic Office as Admiralty Chart No. 524 in 1823.

Appendix

Published Charts Partially Credited to George Gauld

A Chart of the Gulf of Florida or New Bahama Channel, commonly called The Gulf Passage, between Florida, the Isle of Cuba, & the Bahama Islands: From the Journals, Observations and Draughts of Mr. Chas. Roberts, Master in the Rl. Navy, Compared with the Surveys of Mr. George Gauld &ca. London: W. Faden, Aug. 1, 1794.

A General Chart of the West Indies and Gulf of Mexico, describing The Gulf and Windward Passages Coasts of Florida, Louisiana and Mexico . . . Drawn from the Surveys taken by Mr. Geo. Gauld, and others, the New Spanish Charts, &c. and adjusted from recent Observations by P. Foss Dessiou, Master of the Royal Navy. Approved by the Chart Committee of the Admiralty. London: W. Faden, June 4, 1808.

Narrative Descriptions

A General Description of the Sea Coasts, Harbours, Lakes, Rivers &c. of the Province of West Florida by George Gauld 1769. [American Philosophical Society Library, Philadelphia, MS 917.59: G23.]

An Account of The Surveys of Florida, &c. With directions for sailing from Jamaica or the West Indies, by the West End of Cuba, and through the Gulph of Florida. To accompany Mr. Gauld's Charts. London: W. Faden, 1790.

Remarks and Directions concerning the Channels and for sailing into Port Royal and Kingston Harbours; with Cautions, to avoid the dangers that lie in the way. London: W. Faden, 1795; 2d ed., 1798.

Observations on the Florida Kays, Reef and Gulf; with directions for sailing along the kays, from Jamaica by the Grand Cayman and the west end of Cuba: also, A Description, with sailing instructions, of the coast of West Florida, between the Bay of Spiritu Santo and Cape Sable. By George Gauld, to accompany his Charts of those coasts, surveyed and published by Order of the Right Honourable the Lords Commissioners of Admiralty. London: W. Faden, 1796.

A New and Enlarged Book of Sailing Directions for Capt. B. Romans', &c. &c. Gulf and Windward Pilot . . . with the additions of Captains W. G. De Brahm, Bishop, Hester, Archibald Dalzel, Esq. George Gauld, Esq. Lieut. Woodriffe, and other experienced navigators. London: Robert Laurie and James Whittle, 1794. [Contains "General Directions for the Dry Tortugas and the Florida Reef and Kays, with their Description, by George Gauld, Esq. Surveyor of the Florida Coast; May 1774, and November 1775."]

G^{eo.} Gauld — Cartographer

Print

A View of Pensacola in West Florida. Vue de Pensacola dans le Floride Occidentale. For the Honourable Sr. William Burnaby Rear Admiral of the Red, Commander in Chief of His Majesty's Ships at Jamaica & in the Gulf of Mexico . . . Geo. Gauld. London: [?].

Index

All references to ships will be found under main entry Ships.

Aberdeen, 2, 11
Academie Royale des Sciences, xvii
Admiralty: and coastal survey, xix, 14, 22; and Gauld, 11–12, 15–16, 21, 43, 72–73, 121, 127, 171, 219–20; and Gauld's charts, 222–25, 227–28, 231
Alabama River, 102
Alafia River, 51, 137
American Philosophical Society, 2, 129–31, 194, 199
Anclote Island, 136–37
Anderson, Walter, 136–39
Anna Maria Key, 48
Antonio, Cape, 23, 149, 154
Apalachee Bay, 23, 58, 86–87, 118, 122 *n*.12
Apalachicola Bay and River, xvi, 24, 58
Ardbrack, 1
Assembly, General. *See* West Florida, General Assembly
Atchafalaya Bay and River, 177
Atlantic Neptune, 40, 222

Bache, Alexander Dallas, 233–34
Bahama Islands, xv, 155, 168, 170, 172
Bahia Honda, 165
Baker, B., 230
Balize, 80, 124, 128, 163, 184
Ballast Point, 51
Banffshire, 1
Baptiste (settler), 100
Barbados Island, 17, 20
Barreda, Frey Blas de, 45
Barret, Joseph, 9
Barton, Captain, 230

Baton Rouge, 213
Bay, Elihu Hall, 215–17
Bay, John, 216
Bay Minette, 102
Bayou Casotte, 104
Bayou Chico, 31
Bayou Grande, 31
Bayou Saint John, 112
Bayou Texar. *See* Salt River
Bayport, 47
Beacon Hill, 68
Beall (shipwright), 159
Bean Point, 48
Bellefontaine Point, 105
Bell Shoals, 68
Biddle, Owen, 130
Big Lagoon, 31, 40, 92
Biloxi, 105–7
Black's Island, 71
Blackwater River, 38–39, 41
Blaise, Cape (San Blas), 24, 47, 58, 70–71
Blakeley, 102
Blue Mountains, 130, 141, 146
Board of Trade. *See* Trade, Lords Commissioners for
Boca Chica (Sambo Key), 167
Boca Ciega Bay, 50
Boca Grande Key (Sara Gold Key), 152, 158
Boggy Bayou, 63
Bolivar peninsula, 179
Bonfouca River, 112
Bon Secour River, 99–100
Borgne, Lake, 112, 123, 180
Botriphnie, 1
Breton Islands, 110, 180
Bromedge, Hugh, 167

243

Index

Brown, Thomas, 18
Browne, Montfort, 94, 96, 115–20, 125–26, 161, 168–70, 190
Bruce, James, 84
Burdon, George, 171, 173–75, 180–81, 185
Burnaby, William, 22, 29, 34, 39–40, 43–46, 56, 58–59, 141, 186–87
Burnaby, William Chaloner, 22
Burnaby Island (Passage Key), 48–49, 56
Bushy Key, 152, 156
Butcherpen Point, 32
Bute, earl of, 57
Buttermire, John, 215–17

Calcasieu River, 177
Calumet, Isle au, 176, 179
Campbell, George, 1
Campbell, James, 160
Campbell, John, 203, 206, 211, 217–19
Campbell (shipowner), 178
Campbell Town, 38, 41, 117
Canaveral, Cape, xv, 172
Careenage (at Deer Point), 34, 40, 114, 135, 182–83
Carkett, Robert, 40, 59–60
Carr, Samuel, 74–75
Cartography, 14, 141
Carysfort Reef, 234
Cat Island, 59, 106, 108
Cedar Key, 133
Cedar Point, 97
Celi, Francisco Maria, 45, 51, 54
Chagrin Point, 93
Chambers, Francis, 19
Chandeleur Islands, 109–10, 172, 214
Charlevoix, Father, 97–98
Charlotte Harbor, 134, 137–38
Chef Menteur, 111
Chester, Peter, 135–36, 159–64, 173, 182, 190–92, 194–95, 203, 206–7, 209–11, 213, 217
Chicoanche River. *See* Sabine River
Choctawhatchee Bay. *See* Santa Rosa Bay

Choctawhatchee River (Pea Creek), 62–63
Choctaw Point, 101
Christian, Robert, 60
Chronometers, xvii–xviii, 17, 19–21, 83, 122
Chrystie, Adam, 206–7, 211
Clarence, William duke of, 227–28
Clark, Daniel, 130
Clarke, Captain, 154
Cobb, Charles, 159, 165–67, 171, 173–74
Cobban, George, 1
Collins, Grenville, *Coasting Pilot*, xix
Coloradoes, 149, 154
Colville, Admiral, xix–xx
Cook, James (Captain), 15 *n*.3
Cook, James (Master), 46–47, 54–55
Coosa River, 102–3
Coral, 145, 158
Corrientes, Cape, 23
Cotton, Rowland, 39–40, 43–44, 46, 55
Coussa Indians, 63
Coussata Indians, 62
Cox, David, 110
Cox's Lagoon, 40
Crombie, James, 34
Crozer, John, 190
Cuba, 23, 149, 227
Cumberland, William duke of, 5
Cumming (watchmaker), 83
Cummings Point, 93
Cuxhaven, 5

Daphne, 100
Dartmouth, earl of, 161–62, 164, 169
Dauphin Island, 74, 76, 96–98, 105, 218
Davey, Thomas, 174
Davis Point, 65
Deans, Robert, 214–18
DeBrahm, William G., xix, 37, 194, 198, 227, 230
Deer Point, 32, 34–35, 40, 135
Deer River, 100–101
Des Barres, J. F. W., xix, 40–41, 222

244

Index

Dessiou, P. Foss, 231
Devil Point, 41
Dickson, Alexander, 161-62, 180, 203, 212-13
Dog Island, 107, 113
Dog River, 101
Dolland (instrument-maker), 83
Doyle, John, 58
Drummond, James F. E., 139
Dry Tortugas. *See* Tortugas
Dullen River, 1, 92
Dupont (settler), 100
Durnford, Elias, 33-35, 37, 94, 100, 126-27, 136, 160-61, 187, 192-93, 211, 217
Durrell, Captain, 86

Eagle Harbor, 71
Earthquakes, 8, 145
East Bay, 41
Eastertown, 1
East Florida, xvi
East Key, 142
East Lagoon, 40
East Point, 32, 41
Egmont, earl of, 56, 82
Egmont Island, 47, 49, 52-53, 56, 58, 134
Eleven Mile Creek, 92
Eliot, John, 12 *n.*11, 117-20, 140
Ellicott, Andrew, 232
Ellis, John, 195
Emanuel Point. *See* English Point
English Point, 30, 32, 41
English Turn, 124-25
Escambia Bay and River, 30, 38, 41
Espiritu Santo (Tampa) Bay, 40, 43-45, 58, 61, 134, 137, 172; description, 47-54; Gauld's survey, 43-57; Spanish surveys, 44-45
Evans, John, 4-11, 13
Faden, William, xi, xix, 3, 169, 223-31
Fairhope, 100
Fair Point, 32, 34
Falconer, John, 224 *n.*5
Fan Point, 41

Farmar, Robert, 85
Ferguson, James, 129
Fergusson, John, 184-85, 205, 207
Fiddich River, 1, 92
Fishermen, Spanish, 138, 152, 158
Fish River, 100
Fitzherbert, Thomas, 73, 80, 82, 86, 122 *n.*12
Florida, xvi, xix-xx. *See also* East Florida; West Florida
Florida, Cape, 128, 143, 155, 159, 167
Forbes, John, 83-84
Forrest, Arthur, 12-13, 118, 128-29, 141
Fort Augusta, 140, 145-46
Fort Bute (Manchac), 161-63, 203, 205-7, 212-14
Fort Charles, 140
Fort Charlotte, 102, 203, 217
Fort Chartres, 198
Fort Dade, 50, 53
Fort Saint Marks. *See* Saint Marks, Fort
Fort Walton, 62
Fothergill, John, 129
Four Mile Point, 62
Fowl River, 98, 100
France, xvi-xviii
Franco, Juan Baptista, 45
Francois (settler), 100
Franklin, Benjamin, 129, 199
Funchal, 19

Gadsden Point, 51
Gage Hill, 213, 219
Galveston Bay, xv, 80, 179
Gálvez, Bernardo de, 123, 179-80, 204-5, 212-13, 217-19
Garcon Point, 41
Garrow, Joseph, 84
Gauld, Ann (wife), 192-93, 219, 223-26, 228
Gauld, George
—life: birth and youth, 1-2; at King's College, 2, 11; naval schoolmaster, 2-4, 11-12; in North Sea, 4-5; in Mediterranean, 5-10; with Forrest,

245

Index

11–12; appointed surveyor, 14–16; voyage to Pensacola, 15–24; visits Havana, 80–81; observes transit of Venus, 120–21; at New Orleans, 124–25; and American Philosophical Society, 129–31; work interrupted by war, 182–85; conditions of work, 186–88, 221; marries, 192–93; and Bernard Romans, 194–98; and Thomas Hutchins, 198–201; and defense of Pensacola, 213, 218–19; as naval agent, 214–17; to New York, 219; to London, 219; death, 220; will, 223
—properties: at Pensacola, 33–35, 219; on Pearl River, 123; on Thompson's Creek, 163; 190–92
—public service: 118, 188–90, 214; in Assembly, 115–17, 119, 127–28, 206–11
—surveys: Pensacola Bay, 29, 38; Espiritu Santo, 40, 43–57; east of Pensacola, 60–66, 68–72; with Pittman, 86–91; Perdido Bay, 91–94; Dauphin Island, 96–98; Mobile Bay, 96, 99–103; Mississippi Sound, 103–13, 123; the lakes, 123–25; Mississippi River, 128–29, 161–63; East Florida coast, 133–38; Port Royal, 143–47; Grand Cayman, 148–49; Tortugas, 149–51, 166; Key West, 151–52, 166; west of the Mississippi River, 175–79
—works: publication of, 82, 155, 222–32; "A Plan of the Harbour of Pensacola" (1764), 29–32, 38, 41, 44, 222; "A Survey of the Bay of Pensacola" (1766), 37–38, 41; "A Plan of the Bays of Pensacola and Mobile" (1768), 91, 122; *Chart of the Bay and Harbour of Pensacola*, 35, 40–41, 223; "Espiritu Santo," 37, 56–57, 187 *n.*1, 196, 227–28; "A Survey of the Coast of West Florida," 81–82; "Sketch of the Entrance from the Sea to Apalachy," 88–89; "A General Description of West Florida," 126, 130, 132, 199–201; "Measurement of Blue Mountain," 130; "A General Plan of Port Royal," 144–46, 155, 222, 229; "The Island of Grand Cayman," 148, 155; "A Plan of the Tortugas and Kays," 155–58, 165 *n.*15, 225; "A Plan of Manchac," 163, 213; "A Plan of the Mouths of the Mississippi," 164; *An Account of The Surveys*, 165 *n.*15, 169, 218, 225–27, 229–30; *An Accurate Chart of the Tortugas*, 169, 225–26, 232; *A View of Pensacola*, 186–87; "A map of West Florida" (Romans), 196; "Remarks on the Tortugas," 199–200, 222; *The Grand Caymans*, 226, 231; *The West End of Cuba*, 226; "A General Chart," 226–27; *Chart of the Gulf of Florida*, 229, 231; *Remarks on Port Royal*, 229; *Observations on the Florida Kays*, 229–30; *An Accurate Chart of West Florida and Louisiana*, 230, 232; *General Chart of the West Indies*, 231

Gauld, Isabel (sister), 223
Gauld, Janet Moir (mother), 1
Gauld, William (father), 1
Gautrais (settler), 123
Gayton, Clark, 165, 167, 170–71, 182
Germain, George, 219
Gibbs, Anthony, 125, 132–35, 149
Gibraltar, 5–6, 10
Gillori Island, 96–98
Gomez, Martinico, 55
Gordon, Adam, 23–25, 27
Gordon, Harry, 198–99
Gosling, William, 215
Gradinego, John, 100
Grand Bature, 104–6
Grand Bay, 104
Grand Cayman, 23, 148–49, 154
Grand Gosiers, 110
Grant, William, 168 *n.*20
Grant's Point, 48
Graveline Bayou, 105
Great Point Clear, 100
Green Island. *See* Verte, Isle
Gulf Breeze, 32, 35, 190

246

Index

Gulfport, 108
Gun Bluff, 148

Haldimand, Frederick, 85, 115, 188, 195, 198, 204
Harmond, Thomas, 74–75
Harrison, John, xviii, 17, 19
Harrison, William, 17, 19–21
Hartley, Richard, 190
Havana, xvi, 80–81, 187 n.1, 217, 219
Hawke's Bay, 96, 98–99
Hay, George, 223
Hellshire, 146
Hernandez Point, 141
Heron Bay, 97
Heron Island, 97, 111
Hillsborough, earl of, 56, 126
Hillsborough Bay and River, 51, 56, 137
Hogstie Bay, 148–49
Hooker's Point, 51
Horn Island, 105–7
Houmas River, 112, 162
Hunt's Bay, 140, 143
Hurricanes, 74–79, 208
Hutcheson, Francis, 195
Hutchins, Anthony, 206–8
Hutchins, Thomas, 130, 161, 163, 169, 198–201

Iberville River, xvi, 112, 128, 160–63, 213–14
Indian Bayou, 41
Indian Key, 234
Indians, 26, 62, 103, 105, 111, 123, 177–79
Instruments, surveying, 15, 21, 73, 83, 121–22, 151–52, 169–70, 172, 224. *See also* Chronometers
Iskenderun, 9–10
Isla River, 1, 92

Jackson, William, 133
Jamaica, 22–23, 130, 140–48; Assembly, 225–26, 228
Jefferys, Thomas, 187, 223, 227–28
John's Pass, 50
Johnston (artillery captain), 216

Johnstone, George: governor of West Florida, xvi, 12–14, 16, 27; at Pensacola, 28–29, 32–36; on Gauld, 36–37, 57, 67, 79, 85, 119 n.6, 140, 190
Jones, Evan, 195
Jungle Estates, 50
Juniper Creek, 63
Jupiter, xviii, 120

Keith, Basil, 86, 95
Kelly, Timothy, 214
Kennedy, Captain, 121
Key Biscayne, 234
Key Largo, xv, 167, 172
Key West, 150–52, 154, 158, 165, 167–68
Keys, Florida, 128, 134–35, 165, 167, 172, 186, 233–34
Khios Roads, 8–9
King's College, 1–2, 11
Kingston, 22–23, 140–41, 143, 145. *See also* Port Royal
Kirkland, James, 183–84
Knatchbull, Charles, 142–43, 146
Krebs, Hugo, 104

La Combe River, 112
La Desirade Island, 21
La Ronde, M., 123
Le Faber, 123
Leghorn, 7
Le Montais, Francis, 170
Lindegren, Nathaniel, 169
Lindsay, John, 14–25, 28–29, 32, 39, 43–44, 55, 58–59, 73, 83, 121
Little Dauphin Island. *See* Gillori Island
Little Manatee River, 51
Livingston, Phillip, 160, 192, 195
Lloyd, Thomas, 180, 182
Longboat Pass, 50
Long Island (Anna Maria Key), 48
Longitude: determining, xvii–xviii; Board of, xviii
Lopez, Juan, 227
Lora Point, 41
Lorimer, John, 116, 120–21, 130, 132,

Index

161–63, 169–70, 190, 193–98, 201, 219, 223–25, 227
Louisiana: ceded to Spain, xvi; survey of coast, 175–79

McGillivray, 101
McNemara, James, 214–16
McPherson, James, 187
Madeira, 18–19
Magnolia River, 100
Maistell, Thomas, 34
Malhereux Islands, 111
Mallory, Stephen R., 234
Manatee River, 51
Manchac. *See* Fort Bute
Marquesa Islands, 135, 152, 157
Marsh Island, 177
Massacre Island. *See* Petit Bois Island
Matacumbe, 167
Maurepas, Lake, xvi, 112, 123, 160–62
Maxent's Bayou, 123
Mease, Edward, 39
Mediterranean Sea, 5–10
Mermentau River, 177
Messina, Strait of, 7–8
Middle Key, 150
Middle Knoll, 141, 145
Mikisuki, 87, 90
Miller, John, 207–9
Mississippi River, 124–25, 160–63, 168–69, 179, 184, 203; mouths of, 80, 124–25, 128–29, 144, 163–64, 172, 175–76, 196–97, 199
Mississippi Sound, 103–9
Mitchel, Elspet, 1
Mitchell, John, 211, 213, 215–17, 219, 224 *n.*5
Mobile, xvi, 25, 101–2, 190, 203; representation of, 206–10; road to, 32, 93–94, 120; seized by Spain, 217–18
Mobile Bar, 74, 76, 99
Mobile Bay, 96, 98–103
Mobile Point, 98–99
Mobile River, 101–2
Mole Saint Nicholas, 143

Monberaut, Montault de, 100
Mon Louis, Island (Isle aux Maraguans), 100
Montalba, Lorenzo de, 45
Moore, Alexander, 224 *n.*5
Morrison (lawyer), 216
Mulatto Bayou, 41
Mullet Kays, 47, 49–50, 53, 55–56
Mullet Point, 100
Murray, George, 59, 66–67, 70, 73–79

Nairne, Edward, 129
Nassau Road, 110–11, 173
Natchez, 162–63, 203, 205, 207–8, 230
Navigation, xvii, 16–17, 22–24
Navy Board, 14–16
Nelson, Horatio, 141
New Orleans, xvi, 39, 112–13, 123–25, 168, 175, 179, 203, 205
New York, 219
New York Daily Advertiser, 226
Nitalbany River, 112
Noble, James, 187
Nollie Creek, 100
North Key, 150
North Sea, 4–5
North Sound, 148–49
Nunn, Joseph, 183–84, 205, 207–8, 230

Ochlockonee River, 90
O'Reilly, Alejandro, 125
Orleans, Isle of, xvi, 128–29, 180. *See also* New Orleans
Osborn, John, 174–75, 178–84
Ostend, 4
Oyster Bay, 100
Oyster Bayou, 177
Oyster Point (Fla.), 32
Oyster Point (La.), 177

Pakenham, John, 121–25
Palisadoes, 140
Palm River, 137
Panama City, 65

248

Index

Parker, Peter, 183–84, 206, 211
Parry, William, 21, 72, 84, 86, 95, 113, 121–22, 128
Pasadena, 50
Pascagoula River, 104–6
Passage Key. *See* Burnaby Island
Pass a La Loutre, 128
Payne, John, 121 *n*.10, 122 *n*.12, 142, 146, 151–52, 154–55, 165–66, 168, 173–79, 192 *n*.6, 229 *n*.19; and *West Florida*, 185, 208, 212–13
Pearl River, 112, 123, 173, 191–92
Pelican Island, 74, 76, 96, 98
Pensacola, xvi; description, 25–27; development, 28, 32–36, 115; Gauld's *View* of, 186–87; hurricane, 208; Spanish attack, 211, 213, 217–19; mentioned, 41, 85, 115, 120, 126, 159, 188-91, 194–95, 206–8
Pensacola Bay, 24–25; Reid's "Remarks," 29; Gauld's description, 29–32; further surveys, 38–41
Perdido Bay and River, 73, 91–94
Petit Bois Island, 97, 105–6
Phillips, Nathaniel, 143, 146–53
Philipps, William, 119–20, 122 *n*.12
Pickles, William, 205, 212
Pilots: at Pensacola and Mobile, 74, 81, 189–90; at mouths of the Mississippi, 80, 164
Pinellas County, 50
Pinto Island, 102
Pittman, Philip, 26, 85–91, 102–3, 160, 163, 193, 198–99
Point au Fer, 177
Point aux Herbes, 104
Pointe aux Chênes Bay, 104
Pollock, Oliver, 180, 204
Pontchartrain, Lake, xvi, 83, 111–12, 123, 175, 212
Portersville Bay, 104
Port Royal, 140–47, 186
Powel (purser), 9
Prevost, Augustin, 26
Privy Council, 14
Proclamation, October 7, 1763, xvi
Providence, 150, 152, 155, 172

Quash, Peter, 84, 95, 113, 119, 153, 165, 173

Raccoon Point, 177
Red Bluff (Perdido Bay), 92
Red Cliffs (Pensacola), 31, 173, 211, 213, 218
Reid, Robert, 18, 23–24, 29, 31
Reid's Tree, 30–31
Riaño, Francisco, 219
Richardson Hammock, 71
Rigolets, 111–12
Riviere aux Poules. *See* Fowl River
Roan, Samuel, 60
Roberts, Charles, 229
Robertson, Archibald, 160
Robertson, James, 26
Robinson, Charles, 173, 184–85
Rochon, Augustine, 93, 100–102
Rochon, Pierre, 101
Rocky Bayou, 63
Rodgers, C. R. P., 233
Rodney, George Brydges, 139, 141, 143–44, 146–48, 154–55, 159, 166, 222, 229
Rodriguez Key, xv
Roebuck Bay, 34
Romano, Cape, 134
Romans, Bernard, 130, 132, 194–98
Ross, David, 117
Round Island, 105–6
Royal Society, 121, 170

Sabine River, 178–79
Sable, Cape, 134
Saint Andrew Bay, 24, 39, 60, 63–67
Saint Augustine, xvi, 85, 88, 90–91, 115
Saint George, Cape and Island, 24, 58, 69
Saint Joseph Bay, 24, 39, 60, 68–71, 133, 136, 153
Saint Joseph Island, 111
Saint Louis Bay, 106, 111
Saint Marks: Fort, 85, 87–89, 91; River, 87–89
Saint Petersburg, 51

Index

Salamis, 10
Salkeld, John, 164
Salt River, 32, 40
San Antonio, Cape. *See* Antonio, Cape
San Blas, Cape. *See* Blaise, Cape
Sand Key (Little Pelican Island), 98–99
Sand Point (Sandy Point), 32–33
Sanibel Island, 134
Santa Rosa Bay (Choctawhatchee Bay), 60, 62–64
Santa Rosa Island, 24, 32; Gauld's plantation, 33–35; careenage, 34, 40; described, 61, 63, 218
Sara Gold Key. *See* Boca Grande Key
Sarasota, Pass, 50
Savage, Gerrald, 212
Scots, in politics, 14, 119
Seahorse Reef, 133
Seller, John, *Atlas Maritimus*, xix
Seven Years War, xv–xvi
Shalimar, 62
Shaw, Farquhar, 223
Shell Keys, 177
Ship Island, 39, 59, 75, 106–9, 112, 121–22, 172–73, 180–81
Ships: *Achilles*, 141, 145; *Active*, 39–40, 43, 59, 114; *Adventure*, 72–73, 79–80, 84–86, 103; *Alarm*, 39–40, 43, 45–47, 50, 53–55, 58; *Atalanta*, 175, 180, 182; *Augusta*, 13; *Augustus Caesar*, 148; *Berwick*, 6; *Betsey*, 45–47, 50, 55, 60; *Boyne*, 142; *Carysfort*, 145; *Catherine*, 184; *Centaur*, 11–12, 128; *Charlotta*, 60, 62, 64–66, 72, 82–83, 121; *Cumberland*, 148; *Cygnet*, 85–86; *Dilligence*, 154, 167, 173–74; *Druid*, 22, 109, 113, 129; *Earl of Bathurst*, 218; *Earl of Northampton*, 121–26, 129, 133–39, 142–43, 146–55, 159; *Egmont*, 143; *El Grand Poder de Dios*, 215–17; *Ferret*, 39–40, 59–60, 66–68, 70–80, 83, 153; *Florida* (small schooner), 147, 149–52, 165–66, 174–75; *Florida* (sloop), 159, 165–68, 170–71, 173–83, 205; *Florida* (schooner), 183, 206; *Hawke* (privateer), 9–10; *Hawke* (survey ship), *see* Sir Edward Hawke; *Hound*, 181–84, 205, 208, 214–15, 218, 230; *Jamaica*, 119–22, 125, 142, 166; *Levant*, 86, 91, 95; *Magdalene*, 10; *Mentor*, 214, 218; *Merlin*, 143; *Morris*, 212; *Nelly*, 215; *Northampton*, *see* Earl of Northampton; *Orpheus*, 6; *Plaidas*, 8; *Portland*, 154, 165; *Port Royal*, 214–15, 218; *Preston*, 2–11; *Prince Edward*, 7, 45, 56; *Princess Amelia*, 142; *Renown*, 113, 122; *Revenge*, 6; *Robert*, 178; *Saint John*, 168, 170; *Sally*, 153, 214; *Savage*, 168; *Sir Edward Hawke*, 95–96, 99, 101, 103, 106, 109–14, 119; *Stag*, 129; *Stork*, 170, 208, 218; *Sylph*, 183–85, 205, 208; *Tartar*, xviii, 14–24, 29, 39, 56, 58–59, 83; *Tickel Bender*, 58; *Tryal*, 119, 122 *n*.12; *Tryton*, 8; *Two Brothers*, 181; *West Florida*, 171, 173–75, 180–82, 185, 205, 212–13; *Whim*, 39; *William and Elizabeth*, 214; *Zephyr*, 143
Siguenza Point, 218
Sisson, 169
Six-Mile Creek, 51
Slaves, 133
Smith, James, 66, 68, 70
Smith, Joseph, 34
Smyrna, 8–9
Soldier Creek, 92
Spain, xvi, xx, 50–51, 204–5, 212; and Gauld's charts, 231–32
Spanish River, 101–2
Spence, Robert, 137
Sprague Point, 88
South, Rev. Mr., 9
Southwest Key, 150–51
Stirling, Lord, 129
Stork, William, 187 *n*.1, 227–28
Stuart, John, 194, 196
Survey, of American coasts, xix–xx. *See also* Gauld, George, surveys
Surveying vessels, described, 59, 82,

250

Index

86, 121, 154, 182. *See also* Ships
Suwannee River, 117

Tait, David, 92
Talbot, George, 119, 121–22
Tallahassee, 87, 90
Tallapoosa River, 102–3
Tampa, name of, 56
Tampa Bay. *See* Espiritu Santo
Tartar Point, 31, 173
Tavernier Key, 167
Terrebonne Bay, 176, 179
Terry, Jeremiah, 100
Texas, 179
Thompson (shipowner), 178
Thompson's Creek, 163, 191
Tickfaw River, 112
Tiger Shoal, 177
Timbalier Bay and Island, 176
Tombecbe River, 102–3
Tortugas, 23, 128, 149–57, 165
Trade, Lords Commissioners for (Board of Trade), xix–xx, 14
Tripp, Thomas, 81
Turtles, 148, 150, 154, 157

Ugarte, Thomas, 231–32
United States Coast Survey, 231, 233–34
Unzaga, Governor, 164

Vaca Key (Cayos Vacas), 167
Valparaiso, 62
Venus, transit of, 120–21
Vermillion Bay, 177
Verte, Isle (Green Island), 164
Vice-admiralty court, 215–17

Vicksburg, 169
Village, The, 93, 100, 218
Vincent, 123

Waccasassa Bay (Amasura), 133
Wakulla River, 87–90
Waldeck Regiment, 203, 206, 211, 218
Walnut Hills, 169, 205
Warburton, Charles, 35 *n*.13, 95–96, 99, 101, 103, 106, 109, 111–13
Ward, Daniel, 100
Watson, William (scholarship), 2
Webley, John, 214
Wedderburn, David, 57
Weeki Wachee River, 47
Weeks's Bay, 100
Wegg, Edmund Rush, 100, 195, 214–17, 224 *n*.5
West Florida, xvi, 128, 174; General Assembly, 93, 115–19, 127–28, 135, 159–60, 206–10; and American Revolution, 173–74, 202–20
West Lagoon, 40
Whitefield, George, 1 *n*.1, 220
White Point, 41
Williamson, Hugh, 130
Willing, James, 164, 205–6
Wilson, John, 17 *n*.9
Wreck Reef, 142
Wright, James, 87, 91
Wyatt, John, 9

Yazoo River, xvi, 162
Yellow River, 38–39, 41
Yemassee Point, 41
Ysasbiribil, Mariano, 232